Photochemistry

Carol E. Wayne

The National Health Service, Bristol

Richard P. Wayne

The Physical and Theoretical Chemistry Laboratory and Christ Church, University of Oxford

OXFORD NEW YORK TOKYO
OXFORD UNIVERSITY PRESS
1996

Oxford University Press, Walton Street, Oxford OX2 6DP
Oxford New York
Athens Auckland Bangkok Bombay
Calcutta Cape Town Dar es Salaam Delhi
Florence Hong Kong Istanbul Karachi
Kuala Lumpur Madras Madrid Melbourne
Mexico City Nairobi Paris Singapore
Taipei Tokyo Toronto
and associated companies in
Berlin Ibadan

Oxford is a trade mark of Oxford University Press

Published in the United States
by Oxford University Press Inc., New York

A catalogue record for this book is available from the British Library

Library of Congress Cataloging in Publication Data
Data available

ISBN 0 19 855886 4

Typeset by the authors
Printed in Great Britain by
The Bath Press, Avon

Series Editor's Foreword

Oxford Chemistry Primers are designed to provide clear and concise introductions to a wide range of topics that may be encountered by chemistry students as they progress from the freshman stage through to graduation. The Physical Chemistry series contains books easily recognized as relating to established fundamental core material that all chemists need to know, as well as books reflecting new directions and research trends in the subject, thereby anticipating (and perhaps encouraging) the evolution of modern undergraduate courses.

In this Physical Chemistry Primer, Richard and Carol Wayne have produced an authoritative and beautifully written account of *Photochemistry* which covers both elementary and advanced aspects of the subject and which will be of lasting and broad value throughout any undergraduate's career. This Primer will be of interest to all students of chemistry (and their mentors).

Richard G. Compton
Physical and Theoretical Chemistry Laboratory, University of Oxford

Preface

When visible or ultraviolet light is absorbed by a molecule, the result may be that the molecule behaves as though it is an entirely new species, with chemical properties that are distinct from those of its parent. Absorption in different wavelength regions may lead to different behaviour. The interaction of light with matter thus gives access to an enormously rich extension of 'dark' chemistry.

The new chemistry is one aspect of the field of photochemistry. Photochemistry also provides some fascinating insights into the way in which chemical reactions occur. But the subject has an importance far beyond the study just of chemistry. Life itself depends on photochemical processes. Photosynthesis, in which energy from the sun is harnessed by living organisms, is an obvious example. Our atmosphere, which supports life and shields us from damaging ultraviolet radiation, has its composition determined by photochemistry. Photochemistry also finds applications in many of Man's endeavours, including photography, photopolymerization, the production of photodegradable polymers, and the synthesis of organic chemicals; an exciting and developing area is the use of photochemistry to provide novel medical treatments. Photochemistry in nature and the applications of photochemistry are topics emphasized in this book in order to give relevance to understanding the principles of our subject.

We wish to thank Pete Biggs for the enormous help he gave us in preparing many of the diagrams for the book. His skilled efforts are much appreciated.

Oxford C.E.W.
July 1995 R.P.W.

Contents

1 Photochemical principles

1.1 The nature of photochemistry

Life on Earth depends, both directly and indirectly, on the influence that light has on chemistry. The energy of the Sun's visible and ultraviolet radiation promotes processes that not only permit the continued existence of life on the planet, but which quite probably led to the development and evolution of life itself. Photosynthesis in plants provides the most obvious example of chemistry driven by light that, at the present stage of evolution, forms a vital link between the utilization of solar energy and the survival of life. The production of carbohydrates that can be used as energy sources by other life forms is one part of the story; the production of oxygen, a major component of our atmosphere, is another.

1.1 The word 'light' has a rather broad interpretation in connection with photochemistry. In principle, it refers to radiation of any part of the spectrum that can promote chemical change. As we shall see shortly, such radiation is generally visible light (roughly 700–400 nm) and radiation of shorter wavelengths in the ultraviolet region. The long-wavelength limit is probably about 2000 nm in the near infrared, while at the short-wavelength side the limit is only a formal one at the start of the X-ray region. In the earlier history of the subject, the short-wavelength limit was set by the transmission characteristics of conventional transparent container materials (glass and quartz) and by absorption by the constituents of the atmosphere itself, although these limitations can be overcome to some extent.

Interactions of light with matter make up the fascinating subject of *photochemistry* that is the subject of this book. The photosynthetic process itself and the photochemistry of the atmosphere will both be discussed in more detail in Chapter 5. For the time being, they serve as examples of the importance of light in chemistry. Photosynthesis *uses* light to bring about chemical change. Both in nature and in the laboratory, energy released in chemical reactions can also *produce* light; fireflies provide an illustration of the *chemiluminescence* from an energetic reaction. The chemical changes that follow the absorption of light and the chemistry that produces light are equally examples of photochemical processes. Although we have emphasized here *chemical* change, a number of physical processes that do not involve any overall alterations in chemical identity are taken to lie within the province of the photochemist. *Fluorescence* and *phosphorescence*, in which light is emitted from a chemical species that has absorbed radiation, are both examples of such processes.

If there is one essential feature of photochemistry, it is probably the way in which *excited* states of atoms or molecules (i.e. those with excess energy) play a part in the processes occurring. We shall return very soon to this topic; note for now that the excited states of most concern to the photochemist are those in which the electrons have become distributed into a higher energy arrangement; the species are thus said to be *electronically excited.* Absorption and emission of radiation to and from excited states is studied in *spectroscopy*, and there is much common ground between spectroscopy and photochemistry, as there also is between photochemistry and quantum theory. The *rates* of photochemical reactions are often of interest, so that the concepts of *reaction kinetics* are frequently invoked in quantitative photochemistry. The intertwining of photochemistry with these other branches of chemistry means that the reader needs to have some acquaintance with the underlying ideas. At the same time, the study of photochemistry provides examples of the

1.2 The lowest, unexcited, electronic state of an atom or molecule is referred to as the *ground* state.

application of the other disciplines, and acts as a link between them. As far as possible, we shall attempt to introduce at the most basic level concepts that may be unfamiliar.

Man has been aware from the earliest times of the influence that the Sun's radiation has on matter and the environment. However, it is mainly during the present century that a systematic understanding of photochemical and photophysical processes has developed. A logical pattern to the interactions between light and matter emerged only after the concept of quantization of energy was established. It is the purpose of this book to clarify this pattern, and to explain the foundations on which modern photochemistry is based.

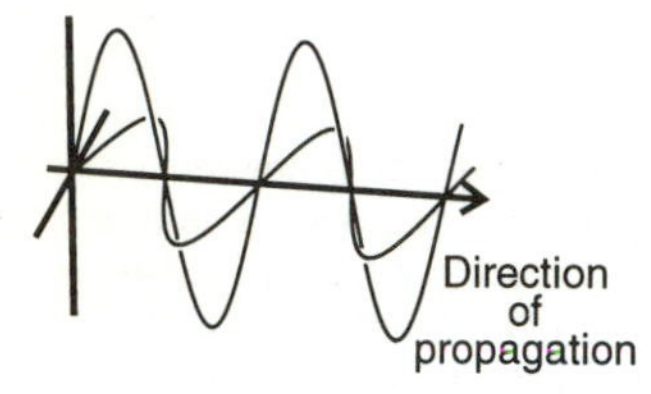

Fig. 1.1 Oscillation of electric and magnetic fields in the propagation of light.

1.2 The nature of light

Light is a form of energy that exhibits both wave-like and particle-like properties. It is propagated in a straight line (unless reflected or refracted) and its speed in a vacuum is approximately $3 \times 10^8\,\mathrm{m\,s^{-1}}$. Light is characterized by its colour, an attribute identified by Newton when he found that a beam of white light could be split into its component colours using a prism.

James Clerk Maxwell showed that light is a form of electromagnetic radiation; its wave-like properties arise from the transverse oscillation of electric and magnetic fields in planes perpendicular to each other and to the direction of propagation of the light (see Fig. 1.1). It is the wavelength or frequency of these oscillations that defines the characteristic colour, i.e. light of a specific wavelength will be a certain colour (red light having a longer wavelength than blue light) [Note 1.3]. The wave theory accounts for many intrinsic properties of light such as propagation, reflection, refraction, diffraction, interference and polarization. There are, however, some phenomena that it is not possible to explain by treating light as a wave; these include the *photoelectric effect* and the nature of *black-body radiation*.

1.3 The wavelength and frequency of electromagnetic radiation are related by the expression

$$c = \nu \times \lambda$$

where c is the velocity of light (expressed in $\mathrm{m\,s^{-1}}$), ν is the frequency (Hz), and λ is the wavelength (m).

Einstein's work on the photoelectric effect demonstrated that radiation also behaves as though it consists of a stream of particles, or *photons*. Each of these photons has a fixed energy that depends on the frequency (or wavelength) of radiation [Note 1.4]. The postulate that light behaves as particles of defined and indivisible energy is closely related to *quantum theory*, which had been developed by Max Planck in connection with his work on black-body radiation. It is not only physical phenomena for which quantum theory provides an explanation. The chemical changes wrought by electromagnetic radiation in its interaction with matter would be very difficult to interpret without assuming that the radiation can behave as particles whose energy is quantized.

1.4 The energy (E) of photons is related to the frequency (ν) of the radiation by the expression

$$E = h \times \nu$$
$$= h \times c/\lambda$$

where h is Planck's constant.

Light quanta may only interact with atoms or molecules one at a time [Note 1.5], so that the energy available to each reacting atom or molecule is the energy possessed by the photon with which it is interacting ($h\nu$) [Note 1.4]. The *intensity* of light is related to the number of photons per unit time; the energy of the individual photons is fixed. An increase in the intensity of radiation will therefore only increase the number of species excited

1.5 If the light is of particularly high intensity (e.g. from a laser), then multiple photon absorption may occur (see Section 2.12).

and not the energy available to each individual species. This characteristic has been demonstrated experimentally: it may be looked at as a chemical equivalent of the photoelectric effect, and is the basis of the *Stark–Einstein* law described in Section 1.8.

1.3 Absorption and emission of radiation

Energy levels in matter are quantized; i.e. species can exist only in certain defined, discrete (separate) energy states. A transition between two such states of specific energy must also have associated with it a definite energy. Thus, a direct result of the quantization of energy levels is that for each individual species only specific energies, and therefore frequencies, of radiation can be absorbed or emitted. The characteristic line and band spectra of chemical species are a consequence of this behaviour.

1.6 Other interactions, with electric quadrupoles, magnetic dipoles, and magnetic quadrupoles, are also possible, but they tend to be weaker than the electric dipole transitions.

In addition to its translational energy, a species may possess other sorts of internal energy, each of which is quantized: rotational energy, vibrational energy which arises from the periodic oscillation of atoms in a molecule, and electronic energy which depends on the distance of the electron from the nucleus and the type of orbital that it occupies.

An atom can have only electronic (and no rotational or vibrational) internal energy; any increase or decrease in energy of the atom must therefore result from a change in the electronic state as the electron moves between the various available orbitals. These electronic transitions also occur in molecules, between the molecular bonding, non-bonding and anti-bonding orbitals that are available to the electron. Although the situation is more complicated for molecules, as they can also undergo transitions between the vibrational and rotational energy levels, it is the electronic transitions which are of most interest to photochemists for the reasons that will be discussed in Sections 1.4 and 2.2.

1.7 Usually, the absorption occurs in a single step, so that the difference in energy between the states *l* and *m* must correspond exactly to the energy of the photons of the incident radiation. This situation is called the *Bohr condition*: absorption can only occur if $\varepsilon_l - \varepsilon_m = h\nu$ (see Fig. 1.2).

When a species absorbs a quantum of radiation, it becomes excited. How the species assimilates that energy — in rotational, vibrational, or electronic modes — depends on the wavelength of the incident radiation. The longer the wavelength of electromagnetic radiation, the lower the energy [Note 1.4]. As the molecular rotational and vibrational energy levels are closer together in energy than the electronic levels, absorption at the lower energies associated with infra-red radiation usually only leads to transitions between rotational or vibrational states of the species. The energy of radiation in the visible to ultraviolet region, however, is of the right magnitude to induce transitions between the electronic energy levels of a species, leading to electronic excitation. This interaction usually takes place via the electric dipole (permanent or transition) of the species [Note 1.6].

In order to discuss different types of absorption and emission processes, we use a model in which a chemical species possesses two quantized states *l* and *m*, of energies ε_l and ε_m, where *l* is at a lower energy than *m*. To reach state *m*, a species in state *l* must gain energy; it does this by absorbing energy from the electromagnetic radiation [Note 1.7 and Fig. 1.2].

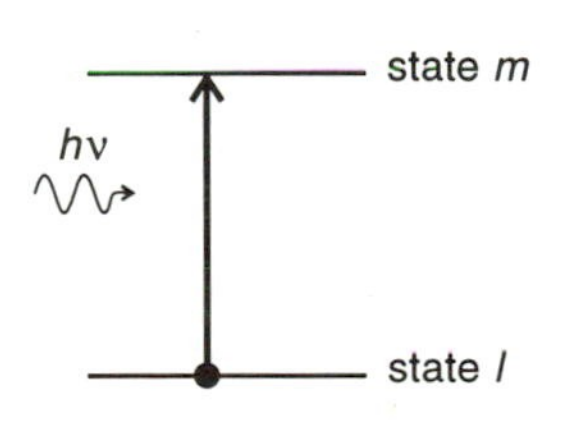

Fig. 1.2 Excitation by absorption of radiation.

The converse of absorption, when a chemical species undergoes a transition from the higher energy state l to the lower energy state m, constitutes emission. There are, in fact, two types of emission. *Stimulated emission* is the exact analogue of absorption; the other process is *spontaneous emission.* In the case of stimulated emission, the interaction between the excited species and the electromagnetic radiation causes the energy-rich species to give up its energy to the incident radiation [Note 1.8]. In the case of spontaneous emission, the energy-rich species loses its energy of its own accord in the absence of a radiation field. As the emission phenomena considered in photochemistry — namely fluorescence, phosphorescence and chemiluminescence — are normally spontaneous, when emission is referred to in this text from this point onwards we mean spontaneous emission, unless we specifically state otherwise.

1.8 *Lasers* (see Sections 3.7 and 6.8) depend on stimulated emission for their operation.

There is a distinct difference between stimulated and spontaneous emission. It is evident that the rate (or intensity) of either absorption or stimulated emission must be proportional to the rate of collision between photons and atoms or molecules of the absorbing or emitting species, and hence must be proportional both to the radiation density and to the concentration of the chemical species. Spontaneous emission, however, is a random process and is therefore kinetically first order (i.e. it depends only on the concentration of the excited species). The rate constant for spontaneous emission for an excited species is the *Einstein A Factor* [Note 1.9], which can thus be used to give a good indication of the lifetime of the species in question; a smaller rate constant will result in a longer lifetime. It can be shown that, other factors being equal, the efficiency (and hence the rate) of the spontaneous process is proportional to ν^3.

1.9 The *A* Factor determines the intensity of emission through the rate equation

$$I = A[X]$$

where I is the emitted intensity and [X] the concentration of excited species. The concentration (and thus intensity) halves in a period $\tau_{1/2}$, the *half-life*, given by the equation

$$\tau_{1/2} = (\ln 2)/A$$

Another characteristic *lifetime*, τ, is more simply defined as $1/A$ (see pp. 64–65).

The Beer–Lambert law

Absorption is treated quantitatively using the Beer–Lambert law. This law gives the fraction of monochromatic light transmitted through an absorbing system, and is expressed through the relation

$$\frac{I_t}{I_o} = 10^{-\varepsilon Cd} = e^{-\alpha Cd} \tag{1.1}$$

where I_t and I_o are transmitted and incident light intensities, C is the concentration of absorber, and d is the depth of the absorber through which the light beam has passed. ε in the first exponential expression is the *decadic absorption coefficient*; the equivalent constant, α, in the expression using base e is the *natural absorption coefficient.* The relationship follows naturally from the assumption that the rate of loss of photons is proportional to the rate of bimolecular collisions between photons and the absorbing species. The Beer–Lambert law is of particular importance in determining the intensity of absorbed radiation in photochemical experimentation, and in calculating concentrations from absorption measurements.

1.10 The Beer–Lambert law only holds for certain conditions, for instance if the species does not associate or dissociate, and if ε is constant over the range of wavelengths being studied.

1.4 Electronic excitation

The excited state of a species can exhibit a dramatically altered reactivity from the ground state. Not only does the species possess more energy, but it can also participate in different reactions as a result of the new electronic arrangement.

The excess energy of an excited species can alter its reactivity, and is particularly significant in the case of electronic excitation because of the energies involved. First, the energies are of the same order of magnitude as bond energies, so that electronic excitation can have a considerable effect on the bonds in a species. Secondly, the energies correspond roughly with typical activation energies for reactions, and the excitation energy can help a species to either partially or completely overcome an activation barrier (e.g. in situations where reaction of the ground state is highly improbable because the barrier is too high). It can therefore be seen that an energy-rich electronically excited species is likely to differ in reactivity from the ground state, the excited state normally being the more reactive of the two.

1.11 In a molecule, an electron can be promoted to an antibonding orbital, thus making the species more unstable (and possibly leading to dissociation), or it can be promoted into a more strongly bonding orbital and consequently stabilize the species.

The excess energy possessed by an excited species is not the only factor that affects its reactivity. Even if the energy factor had no influence, the excited state could still exhibit a reactivity significantly different from that of the ground state. On excitation, an electron is frequently promoted to a different type (and therefore shape or symmetry) of orbital (e.g. s or σ electrons may be promoted to p or π orbitals). The arrangement of the electrons in a species is fundamental in defining what reactions the species can and cannot undergo, so that such a change in the spatial distribution of electrons can have a major effect on reactivity.

1.5 Wave mechanics and quantum numbers

The quantitative treatment of electronic excitation and the absorption and emission of radiation by chemical species falls within the realm of wave mechanics. This subject is a highly elegant but complex mathematical theory that replaces the classical ideas of position and trajectory with the concepts of *amplitude* and *wavefunction*. The theory gives rise to the quantum numbers that define the electronic state of a species [Note 1.12]. It is frequently the spin, orbital momentum and symmetry properties of the two wavefunctions in upper and lower states that determine whether or not an interaction with electromagnetic radiation is possible; we shall therefore concentrate on the quantum numbers that describe these properties.

1.12 The tools for obtaining the quantum number are based on various forms of the *Schrödinger equation*. For further information, see, for example, P.W. Atkins, *Physical Chemistry (5th ed.)*, OUP 1994, pp.370–382.

Individual electrons possess both spin angular momentum and orbital angular momentum. These momenta are vector quantities; there are therefore angular momenta associated with a species resulting from the vector addition of the spin and orbital angular momenta, a process known as coupling.

The spin moment of a single electron, **s**, is equal to ½. The vector summation of spin moments of each individual electron in a species (*spin coupling*), gives a resultant overall spin **S** and is extremely important in defining the electronic state of a species. It is common practice to specify a state by the *spin multiplicity*, $2\mathbf{S}+1$, rather than by the spin itself [Note 1.13]. For species in which all the electrons are paired, $\mathbf{S}=0$, $2\mathbf{S}+1=1$ and the species is a *singlet*. Most chemical species that can be stored for long periods are of this type, although there are exceptions. If two electrons are unpaired with parallel spins, $\mathbf{S}=1$, $2\mathbf{S}+1=3$ and the species is in a *triplet* state (the excited states of species with singlet ground states are commonly triplet states). Species such as free radicals have one unpaired electron, so that $\mathbf{S}=½$, $2\mathbf{S}+1=2$ and the species is in a *doublet* state. One simple scheme for describing states consists of a numbered list based on the multiplicity where, for instance, the ground-state singlet is denoted S_0 and the excited singlets S_1, S_2, etc., in order of their energies, and the triplet states are T_1, T_2, etc. (T_0 being excluded because it is not the ground state). Spin is of particular significance because, according to quantum mechanics, it cannot change in either radiative or radiationless transitions; states of the same chemical species, but possessing different multiplicities, tend to behave as distinct groups unlikely to convert from one to the other. It is important to note, however, that spin is not always a pure quantity, so that transitions between states with different **S** may sometimes be observed although they tend to be significantly less probable than transitions between states of the same multiplicity (see end of section 1.6).

In atoms, in a simple scheme of coupling angular momenta known as *Russell–Saunders coupling*, the individual spin moments ($\mathbf{s}_1$, $\mathbf{s}_2$, etc.) couple to give **S**, while the individual orbital angular momenta (denoted $\mathbf{l}_1$, $\mathbf{l}_2$, etc.) couple to give a resultant overall angular momentum **L** [Note 1.14].

In diatomic and linear molecules, the electronic state may still be defined in part by the orbital electronic angular momentum. The individual and total orbital angular momenta are denoted λ and Λ respectively, and are used to derive the spectroscopic *term symbol* that describes the state of a species. However, additional information is also required to describe the state completely, including the total angular momentum and also information about the effect that symmetry operations have on the sign of the wavefunction. In particular, whether the wavefunction stays the same, or changes sign, on inversion through a centre of symmetry (if it exists) defines a state as 'even' or 'odd', denoted g or u, respectively; whether the wavefunction stays the same or changes sign on reflection by a plane of symmetry passing through the molecule defines a state as '+' or '–'. For small, non-linear molecules it is still possible to describe the electronic state in terms of molecular symmetry, although it may no longer be possible to specify the state in terms of orbital angular momentum. In the case of molecules that are more complex still, symmetry elements may also be non-existent; the state may then be described in terms of the individual molecular orbitals (e.g. (n,π^*), (π,π^*), (n,σ^*), (π,σ^*) etc. [Note 1.15]) with an indication of the multiplicity included in the term symbol. Spectroscopic transitions are referred to as $n\rightarrow\pi^*$, and so on.

1.13 The reason that the multiplicity, $2\mathbf{S}+1$, is used, rather than **S**, is because $2\mathbf{S}+1$ shows the number of energy sub-levels that can be present. This point is explained in Note 1.14.

1.14 Since **S** and **L** are both vector quantities, they can also combine to give a resultant *total angular momentum*, **J**. There are $2\mathbf{S}+1$ values of **J** (if $\mathbf{L}>\mathbf{S}$), so that the multiplicity provides an indication of how many **J** subcomponents might exist for a particular state.

1.15 The terms (n,π^*), (π,π^*) etc. are derived from the individual orbitals of the molecule: σ and π denote bonding orbitals, σ^* and π^* denote antibonding orbitals, and n denotes a non-bonding orbital. The first character in the term refers to the type of orbital from which the electron was promoted to achieve the excited state (one of the highest filled orbitals of the ground-state molecule); the second refers to the orbital which the electron has entered (one of the lowest unfilled orbitals of the ground state). The multiplicity is shown as a prefix, e.g. $^1(n,\pi^*)$ or $S(n,\pi^*)$.

In the same way that a species has a spectrum resulting from electronic transitions, so are there spectra arising from transitions between vibrational and rotational energy levels. These levels are described by their own quantum numbers. Fig. 1.3 shows the potential energy diagram for an anharmonic diatomic vibrator (the simplest realistic model), with the vibrational levels and probability distributions (the squares of the wavefunctions) also represented. It is assumed for simplicity that electronic, vibrational and rotational energy are entirely independent because of the great differences in frequency between the three types of motion (the *Born–Oppenheimer approximation*). A further consequence of this assumption forms the basis of the *Franck–Condon principle*. The idea behind the principle is that the nuclei of a molecule can be assumed to be fixed during an electronic transition, so that the transition can be represented on a potential energy diagram as a vertical line between upper and lower states as illustrated in Fig. 1.4.

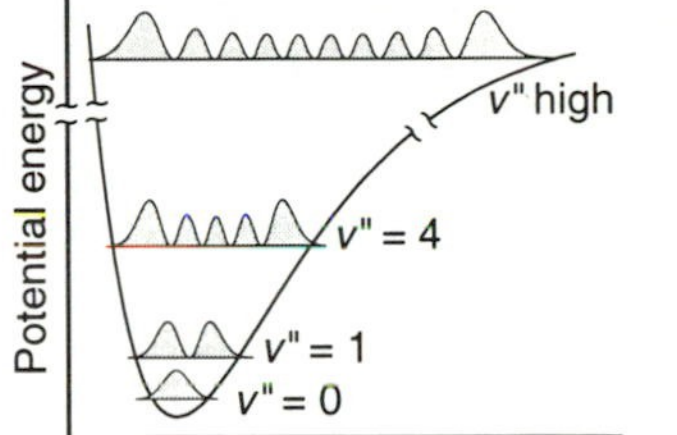

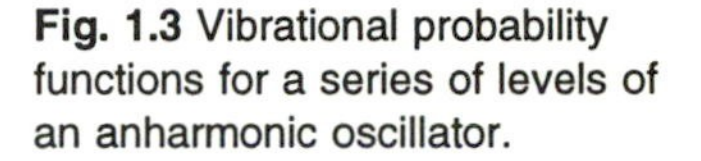
Fig. 1.3 Vibrational probability functions for a series of levels of an anharmonic oscillator.

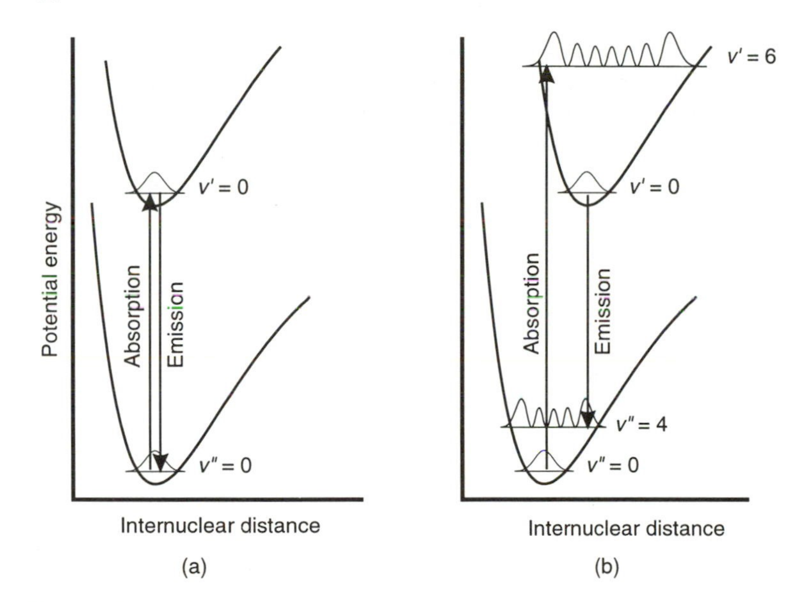

Fig. 1.4 Electronic transitions of greatest probability for absorption and for emission from lowest vibrational levels: (a) where both electronic states are of similar sizes, (b) where the upper state is larger than the lower.

1.6 Selection rules for optical absorption

We have already stated (see p. 5) that wave-mechanical reasoning can show whether or not an interaction with electromagnetic radiation is possible. *Selection rules* are a convenient way of summarizing this information, and give a indication of the feasibility of a transition from one state to another in terms of the quantum numbers [Note 1.16]. They give no indication of the absolute intensities of those transitions.

1.16 In atoms, for one-electron transitions, we have the 'electric dipole' selection rules

$\Delta \mathbf{L} = \pm 1$

$\Delta \mathbf{J} = 0, \pm 1$, but $\mathbf{J} = 0 \nleftrightarrow \mathbf{J} = 0$

For diatomic and linear polyatomic molecules, the orbital momentum rule is

$\Delta \Lambda = 0, \pm 1$

and, where applicable, the rules governing symmetry are

$u \leftrightarrow g,\ + \leftrightarrow +,\ - \leftrightarrow -$

For non-linear molecules that still possess some symmetry, selection rules can again be derived on the basis of the symmetry properties.

Perhaps the most important selection rule is that governing spin multiplicity: spin must not change during an electronic transition. This rule is usually written as

$$\Delta S = 0 \tag{1.2}$$

1.17 Interactions can occur between spin angular momentum and orbital angular momentum, resulting in *spin–orbit* coupling, an effect which increases with increasing nuclear charge; one consequence is that **S** is no longer a strictly valid quantum number, so that selection rules based on it can fail. Vibrational and electronic motions interact to give 'vibronic' coupling; rotational and electronic motions also interact but to a much lesser extent because of the greater difference between their frequencies.

Although selection rules show in principle which transitions may or may not be possible, 'forbidden' transitions are often seen, although they tend to be considerably weaker than allowed transitions. A number of reasons exist for the apparent breakdown of these selection rules. For a start, the formal quantum numbers such as **S** are not always a very good description of the state of a species. Although Russell–Saunders coupling is a fairly good description of the situation for light atoms, it tends to fail for heavier ones, and the breakdown of selection rules based on these vector quantities reflects the inadequacy of **S** and **L** as quantum numbers. Other types of coupling exist, which undermines the assumption that excitation may be considered independently in terms of orbital and spin momenta [Note 1.17]. Secondly, selection rules as stated here refer to electric dipole transitions. A transition that is forbidden by these rules may be allowed for electric quadrupole, magnetic dipole, and magnetic quadrupole interactions [Note 1.6]. Finally, selection rules are derived on the basis that the system is unperturbed, and they therefore tend to break down as a result of collisions or the presence of electric or magnetic fields.

1.18 The various processes represented in Fig. 1.5 may be illustrated by the following examples

(i) **Dissociation**:

$NO_2^* \rightarrow NO + O$

This process is often referred to as *photolysis* (from the Greek for 'splitting by light').

(ii) **Chemical reaction**

Intermolecular reaction:

$H^* + HI \rightarrow H_2 + I^*$

(See also the charge-transfer reactions in Section 2.11).

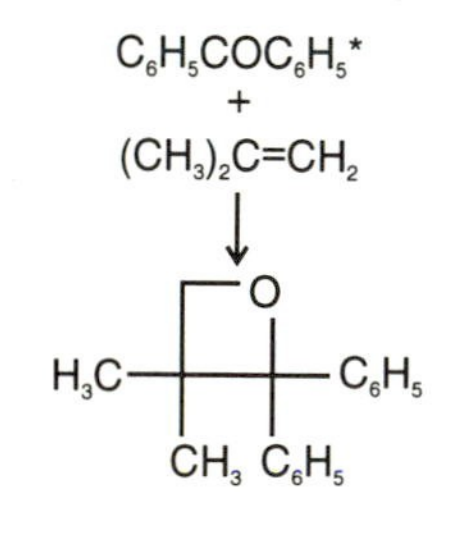

1.7 Fates of excited species

The absorption of radiation by an atom or molecule can lead, as we have seen, to the production of an electronically excited species that can exhibit 'new' chemistry that is different from that of the parent, ground-state species. There are several ways in which the excited species can react or otherwise use or lose its excess energy, and some of the most important routes are represented diagrammatically in Fig. 1.5, with typical examples given in Note 1.18. Each of the routes merits a section to itself, and is considered in more detail in Chapters 2 and 3, although the exchange of energy within a molecule (*intramolecular energy transfer*) is so fundamental to understanding photochemistry that we bring a preliminary discussion of it forward to this section.

The strictly *chemical* routes are (i), (ii), and (iii). The first of these, dissociation, leads to a fragmentation in a molecular species. The energy of the photon has thus been sufficient to rupture a bond. Route (ii) represents the situation in which the excitation is used to promote a reaction (either by overcoming an activation barrier or as a consequence of the new electronic arrangement) that is not possible for the ground electronic state. Although this process is considered as an *intermolecular* process, in which the excited molecule reacts with a separate molecule, *intramolecular* processes are also well known in which one part of an excited molecule attacks another part of the same molecule. This kind of intramolecular process leads to *structural*

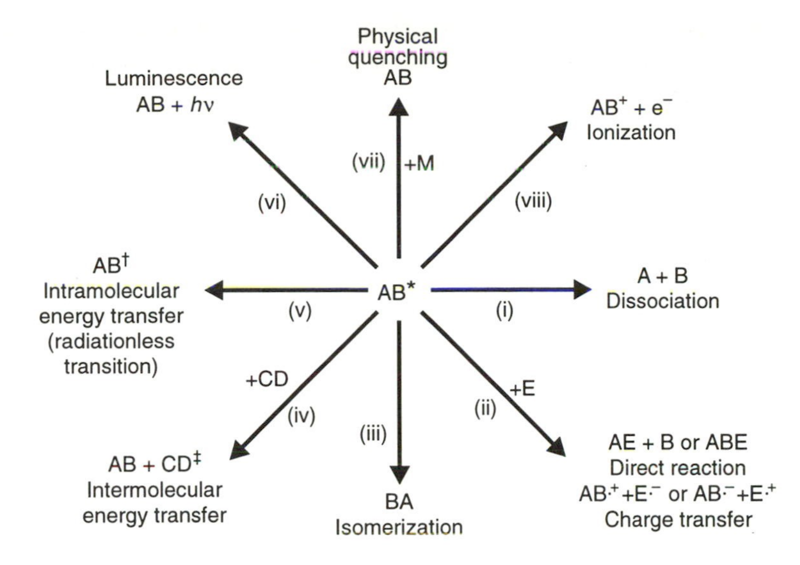

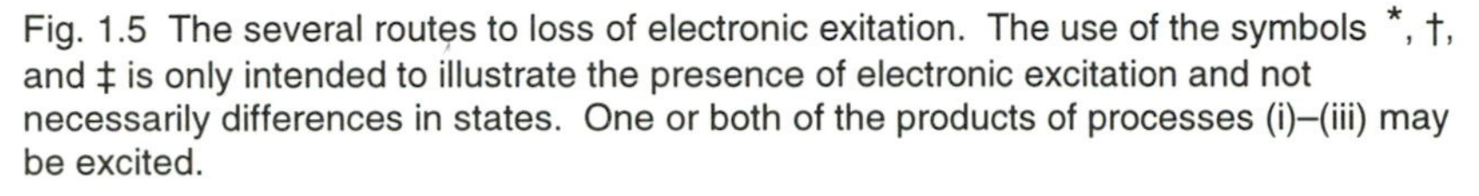
Fig. 1.5 The several routes to loss of electronic exitation. The use of the symbols *, †, and ‡ is only intended to illustrate the presence of electronic excitation and not necessarily differences in states. One or both of the products of processes (i)–(iii) may be excited.

isomerizations or *rearrangements*. Another isomerization process that ought to be included here is channel (iii). An example of this kind of step is afforded by *E–Z* (*cis–trans*) isomerization in which a double bond in the ground state becomes essentially a single bond about which rotation can occur in an excited electronic state [see Note 2.3]. Pathway (viii) is similar in concept to the dissociation step, except that, instead of two chemically distinct neutral fragments being formed, an ion and an electron are produced. Such ionization processes are of considerable importance in many aspects of photochemistry. In the gas phase they are generally of high energy, and require short-wavelength radiation; many of the ions and electrons present in the Earth's atmosphere are formed as a result of *photoionization*.

There are two processes, shown as pathways (iv) and (v), that generate excited electronic states other than the one first formed in the absorption step. These two routes are a collisional *intermolecular* one (pathway (iv)), in which a different molecule becomes excited, and an *intramolecular* one (pathway (v)), in which a new electronic state of the same molecule is populated. Following these *energy transfer* steps, the new excited species or state can lose its excitation in any of the processes in the diagram in which it is able to participate.

Emission of radiation, which corresponds to radiative loss of excitation energy, is shown as pathway (vi). The general term for the process is *luminescence*. Particular aspects of the phenomenon are described as *fluorescence* or *phosphorescence*, as discussed in Sections 3.2 and 3.3

Intramolecular reaction:

$C_3H_7COCH_3{}^*$

↓

$\cdot C_3H_6C(OH)CH_3$

↓

$C_2H_4 + CH_3COCH_3$

(iii) **Isomerization:**

cis-$C_6H_5CH{=}CHC_6H_5{}^*$

↓

trans-$C_6H_5CH{=}CHC_6H_5$

or

dihydrophenanthrene

(iv) **Intermolecular energy transfer:**

$Hg^* + Tl \rightarrow Hg + Tl^*$

$C_6H_6(S_1) + By(S_0)$

$\rightarrow C_6H_6(S_0) + By(S_1)$

By = biacetyl.

See p. 6 for the meaning of S_0, S_1.

(v) **Intramolecular energy transfer:**

$C_6H_6(S_1) \rightsquigarrow C_6H_6(S_0)$

$N(S_1) \rightsquigarrow N(T_1)$

N = naphthalene

See p. 6 for the meaning of T_1.

(vi) **Fluorescence and phosphorescence:**

$Na^* \rightarrow Na + h\nu_{fluor}$

$N(T_1) \rightarrow N(S_0) + h\nu_{phos}$

See p. 6 for the meaning of S_0, T_1.

respectively. If the species AB* has become excited as a result of a chemical reaction, rather than by absorption of radiation, the emission process is called *chemiluminescence*.

(vii) **Physical quenching**:

$O^* + N_2 \rightarrow O + N_2$

(viii) **Ionization**:

$O_2 + h\nu \rightarrow O_2^+ + e$
(See also Note 2.8).

The remaining route in the diagram, pathway (vii), is *physical quenching* or *deactivation.* It is a collisional process in which electronic energy is taken up by an atom or molecule M, and in which the excess energy is converted to translational or vibrational excitation of M. Although it appears to be chemically rather a boring process, ultimately making the electronic energy unavailable for performing 'chemistry', it is a very important one, since most of the other possible routes have to compete with quenching, especially in solution, where collisions are frequent.

The eight reaction routes outlined in this section are *primary* processes. If new products are formed in these reactions, they may then go on to take part in additional *secondary* thermal reactions, which are photochemical only in the sense that the excited species would not have appeared in the absence of light. For example, the fragments formed in a photodissociation step may initiate complex chemical change as well as removing the starting compound.

1.8 The laws of photochemistry; quantum yields

The first law of photochemistry was formulated by Grotthus and Draper at the beginning of the nineteenth century, and states that *only the light absorbed by a molecule can produce photochemical change in the molecule.* The evolution of the quantum theory led Stark and Einstein to develop and modify this law. Initially, they suggested that one, and only one, molecule reacted for each photon absorbed. This hypothesis was subsequently refined to take into account the fact that chemical reaction is only one of a number of competing routes available to an excited species, as discussed in the previous section. The second law of photochemistry thus states that *if a species absorbs radiation, then one particle is excited for each quantum of radiation absorbed* [Note 1.19].

1.19 If a single photon is involved in the interaction, it follows that the photon must have enough energy to promote chemical change in one molecule, as discussed at the end of Section 1.2.

In the previous section, we outlined a number of routes by which an excited species could yield new chemical species as products. For each primary product formed, there may be a number of secondary processes available, so that the overall products of a photochemical reaction can be very diverse. One concept of great value in interpreting the behaviour, to provide information about photochemical reaction mechanisms, is the *quantum yield,* Φ. Φ is *the number of molecules of reactant consumed per photon of light absorbed.* Measured values can range from less than 10^{-6} to more than 10^6, depending on the reaction system; a quantum yield greater than one suggests the occurrence of secondary reactions, and a value greater than two suggests the operation of a chain reaction mechanism [Note 1.20]. However, the quantum yield as defined so far reflects, without distinction, both the efficiency of the primary photochemical process and also the extent of secondary reaction. It is far more useful to consider primary and overall quantum yields, ϕ and Φ, separately.

1.20 Primary and secondary quantum yields are discussed in greater detail in Section 4.3. A secondary step allows another molecule of reactant to be consumed by the product of the primary process, so that Φ could reach 2. If this secondary step also yields a species capable of consuming the reactant, then Φ can exceed 2. This kind of chemistry with sequential secondary steps is a *chain reaction.*

The primary quantum yield should be stated for a specific primary process (i.e. one of the paths shown in Fig. 1.5). In the light of the Stark–Einstein law, it can be seen that the sum of the quantum yields for all primary processes, including deactivation, must be one. If an overall quantum yield is quoted without further specification, it refers to the removal of reactant, although in some cases it is more helpful to quote a quantum yield for the formation of a particular product.

1.9 Intramolecular energy transfer

We have referred several times already to pathway (v) in Fig. 1.5, in which a *radiationless transition* populates an electronic state different from the one that was first excited by absorption. Because this *energy transfer* process occurs within the same molecule, it is an *intramolecular* conversion from one state to another.

Intramolecular energy transfer is of very great importance for a variety of reasons. It may be the dominant primary process for some chemical species, so that the photochemistry of these compounds is essentially determined by the behaviour of the new electronic state. An indirect route via intramolecular energy transfer may populate an electronic state that is inaccessible by direct optical absorption (perhaps because the transition is forbidden and the absorption weak, or because the energy of the excited state means that it does not absorb the wavelengths of radiation incident on the system). One particularly significant example is the population of triplet states, such as T_1 (see p. 6), following initial absorption from S_0 to S_1. Optical absorption from S_0 to T_1 is forbidden by the spin rule, and the absorption is correspondingly weak, so that formation of the T_1 state by direct absorption is very inefficient. As we shall see shortly, the $S_1 \rightsquigarrow T_1$ process is also formally forbidden by a spin-conservation rule; however, it is limited only by the occurrence of other processes (such as quenching or optical emission) that depopulate S_1, since the excited species must eventually lose its excess energy. $S_1 \rightsquigarrow T_1$ conversion may be able to compete successfully with these other loss processes, and thus lead to efficient population of T_1.

1.21 The symbol $\rightsquigarrow$ is used to denote energy transfer processes, rather than the straight arrow ($\rightarrow$) that is used to denote absorption or emission.

Although the examples of the last paragraph have concentrated on cases where there is a change of spin multiplicity, as in the $S_1 \rightsquigarrow T_1$ process, intramolecular energy transfer can also take place between electronic states possessing the same spin. For example $S_1 \rightsquigarrow S_0$ conversion can readily occur with many chemical systems. In fact, the conservation or non-conservation of spin is used as one basis for classification of intramolecular energy transfer, with the conserved process being labelled as *internal conversion (IC)* and the non-conserved one as *intersystem crossing (ISC)*.

Two consequences of the ISC and IC processes are presented at this stage to illustrate the part that they can play in photochemistry. *Phosphorescence* (see Section 3.3) often arises from the T_1 state of molecules whose ground state is a singlet, S_0. Emission of radiation from T_1 follows initial absorption in the singlet system, and subsequent transfer to T_1 by ISC. Another important process involves the dissociation of complex molecules (see Section 2.3). It may be that an electronic state that is reached in an absorption process is strongly bound, and does not fall apart spontaneously, but that some other, lower lying, electronic state might be unstable and lead to formation of fragments. If, then, ISC or IC could populate this dissociative state from the one reached by absorption, fragmentation could follow the absorption of the photon. Both these processes will be discussed in more detail in the appropriate chapter, and are introduced here as illustrations of the rôle played by intramolecular, radiationless, energy transfer.

1.22 The selection rules for radiationless transitions are, where the quantum numbers are appropriate, $\Delta\mathbf{S}=0$, $\Delta\Lambda=0,\pm1$, $\Delta\mathbf{J}=0$, $+\leftrightarrow+$, $-\leftrightarrow-$, $g\leftrightarrow g$ and $u\leftrightarrow u$.

We shall return to a more complete explanation of the factors that determine the efficiency of radiationless energy transfer in Chapter 3. It is, however, useful to summarize some of the main features here. There are two main factors on which attention should be focused at this stage: these are (i) the intrinsic probability of the electronic transition; and (ii) the part played by the shapes, sizes, and energies of the molecules in the two electronic states.

1.23 We shall see in Chapter 3 how the extent of spin–orbit coupling can be affected by various influences internal and external to the molecule. We shall note also, that for molecules described by the (n,π^*), (π,π^*), etc. nomenclature [Note 1.15], the spin-forbidden processes are much more likely between (n,π^*) and (π,π^*) states, in either direction, than between (n,π^*) and (n,π^*), or between (π,π^*) and (π,π^*) states (see Note 3.12 and Section 3.4)

As for optical transitions, whether or not the radiationless transition can occur is embodied in selection rules given in the margin [Note 1.22]. Comparison with the rules set out in Note 1.16 for optical transitions shows that all rules are the same except for $g\leftrightarrow g$ and $u\leftrightarrow u$. As a result, electronic states reached by *allowed* radiationless transitions will have all quantum numbers and symmetry elements, with the exception of g and u, identical to the states reached by direct absorption. For *forbidden* transitions, on the other hand, the quantum numbers may apparently alter, as in going from $\mathbf{S}=0$ (singlet) states to $\mathbf{S}=1$ (triplet) ones. Just as in the case of optical transitions, the occurrence of these formally forbidden transitions reflects the inadequacy of the quantum number to fully describe the situation. One particular source of breakdown of the $\Delta\mathbf{S}=0$ rule is the existence of *spin–orbit* coupling, as described in Note 1.17 for optical transitions [Note 1.23].

The effects of size, shape, and energy of the molecule in its different electronic states can all be rationalized in terms of the Franck–Condon principle (see p. 7), and the essential ideas will be presented in Section 3.4. For the time being, we can illustrate the concept by looking at the potential energy curves for a diatomic molecule; such curves can also be taken as representative cross sections for the potential energy 'surfaces' that describe polyatomic molecules (see p. 20). We show in Fig. 1.6 typical potential energy curves. Absorption of radiation from the X state can populate the upper state labelled B; in the figure, the line drawn within the B curve is intended to show the vibrational level populated, and thus the total of vibrational and electronic energy. The B state is crossed by another one labelled A, and we are interested in intramolecular energy transfer from the B to A states. The first feature to note about the process is that the sum of the vibrational and electronic energies of the initial (B) and final (A) electronic states must be identical: since there are no collisions or fragmentations, no energy can be gained or lost as translation. The energy transfer thus takes place somewhere along the horizontal line set at the energy with which the B state is generated. However, the Franck–Condon principle constrains the electronic transition to

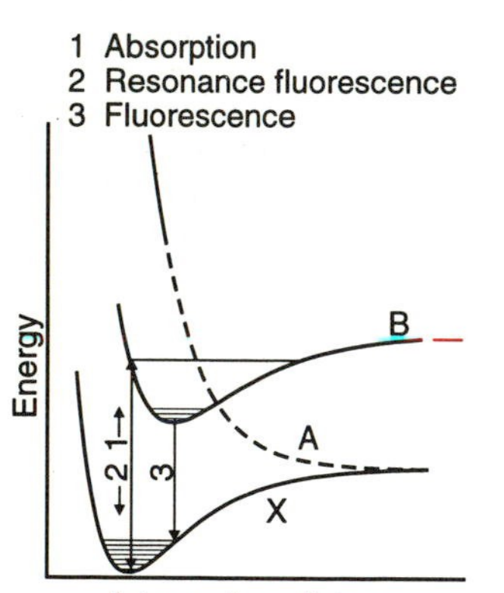

Fig. 1.6 Potential energy curves for the ground state and two excited states of a typical diatomic molecule.

occur without change of nuclear geometry, so that the process must also occur along a vertical line for the nuclear geometry involved. Since neither horizontal nor vertical coordinates can alter, the transfer must occur at a point, although the argument does not yet state at which interatomic distance this point is located. It is often useful to assume that this point where transfer of energy (or crossing from one curve to another) occurs is the point of intersection of the two curves. The fuller argument to be presented in Section 3.4 shows that this is, indeed, the configuration for most efficient energy transfer in most cases, so that the simple picture of jumping from one curve to another at the crossing point is a very helpful one.

For complex polyatomic molecules, it is often not possible to construct the detailed potential energy curves or surfaces. An alternative type of diagram, used for complex molecules such as naphthalene, is presented in Section 3.1, and will be explained there. Such a diagram is called a *Jablonski diagram* after the inventor of its original form.

1.10 Experimental methods

The discussion of experimental techniques presented in this book is only intended to provide a brief overview of the principles involved and of the most common methods used. The nature of the final products of photochemical reaction, and the rates of their formation, can be determined using relatively straightforward standard chemical techniques, which need not be treated here. We concern ourselves largely with those parts of the experimental technique that involve light. Determination of the identities and rates of formation of highly reactive primary and secondary products formed as intermediates can be difficult. Methods adopted to study intermediates include flash photolysis, photolysis in rigid matrices, isotopic labelling, free radical trapping techniques, and use of emission spectra and lifetimes. Precise measurement of the intensity of absorbed (and sometimes emitted) light is essential to the determination of quantum yields, which in themselves are needed in any assessment of the nature and efficiency of primary (and secondary) photochemical processes. It is worth noting at this point that all optical components of the reaction system, particularly the windows of the reaction cell, must be transparent to the wavelength of radiation used for each particular experiment [Note 1.24].

1.24 In addition to the question of transparency of the optical components of an experiment, at short wavelengths ($\lambda \lesssim 190$ nm) air itself begins to absorb significantly. Experiments must then be performed in a vacuum (or with the air flushed out by a non-absorbing gas). For this reason, the uv region at $\lambda < 200$ nm is called the *vacuum ultraviolet*.

Classical experiments

Classical experiments seek to follow the overall course, rate, and quantum yield of a photochemical reaction. The basic idea is to irradiate the reaction mixture with a steady light source, and to identify the products formed. Although the concentrations of reactants decrease with time, and those of the products increase, it is quite common in this type of experiment for the concentrations of the intermediates (e.g. radicals, triplet states, etc.) to remain roughly the same (i.e. at *steady-state* concentrations) for a substantial part of the time of the experiment [Note 1.25].

1.25 See Chapter 4 for a more extended discussion of the steady state.

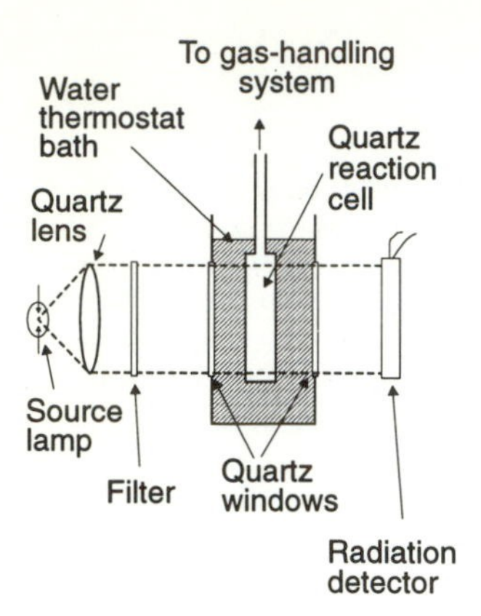

Fig. 1.7 One form of apparatus for 'classical' photochemical experiments using near-ultraviolet radiation.

A typical classical experiment is illustrated in Fig. 1.7. Radiation from the light source passes through a wavelength-selecting device, from which the emergent beam falls upon the reaction cell. Reaction cells are usually constructed with two parallel windows, so that some of the light incident upon the cell may be transmitted through it. A device for measuring radiation intensity is therefore placed beyond the cell to detect any transmitted radiation. The fraction of incident light absorbed can also be calculated, using the Beer–Lambert law (see Section 1.3), from measurements of the incident intensity. Some of the specific components are now discussed in more detail.

Light sources

The 'ideal' light source in a photochemical experiment is monochromatic, to enable precise assessment of the effect of wavelength on the nature and rate of the reaction being studied. However, monochromatic light is not always an experimental possibility — except in the case of laser sources — so that means of isolating narrow wavelength bands must be employed. Techniques used to produce near-monochromatic radiation include grating or prism monochromators, although the transmitted intensity may be so low as to make experimentation impossible; 'colour filters', which may be liquid or solid (glass) solutions containing substances that strongly absorb light of unwanted wavelengths; and 'interference filters' which depend upon the interference effects in thin films (akin to those giving rise to colours in soap bubbles).

1.26 Light sources

Incandescent tungsten-filament lamps have a continuous spectrum which approximates to that of a black body. *Discharge* lamps emit most of their energy in a small number of bands or lines. They can also be used as 'flash lamps'. *Lasers* are extremely well suited to use in photochemical experiments. They provide radiation in a collimated beam which is coherent and monochromatic, is of high intensity, and can be pulsed (see Section 3.7). *Synchrotron* radiation is an extremely effective source of radiation in the vacuum ultraviolet region of the spectrum [Note 1.24], where most other light sources are particularly poor.

Measurement of light intensity

In a photochemical experiment, the amount of light that is absorbed by the reaction mixture must be measured. A light *detector* is used for this purpose. Detectors commonly used include thermopiles, photocells, photodiodes, and photomultipliers.

Thermopiles make use of the heating effect of light as it falls upon a blackened surface; they can be quite sensitive, but are particularly prone to small fluctuations in room temperature and to draughts. They consist of an assembly of thermocouples connected in series, whose front junctions are blackened; the temperature difference between front and rear junctions gives rise to an electrical output (e.m.f) that can be measured.

Photocells are often more useful than thermopiles in the realm of photochemical experimentation, as they are sensitive to light but not to heat. Their purpose also is to convert electromagnetic radiation into an electric current, but directly via the photoelectric effect rather than indirectly via heating effects. A photocell consists of a photocathode and a collector (anode) enclosed in an evacuated bulb; illumination of the cathode causes electrons to be released, and if the collector is positively charged then a current will flow in the external circuit. Operating conditions can be chosen so that this current is proportional to the light intensity reaching the cathode. It is necessary to calibrate photocells against thermopiles or secondary standards (such as the chemical actinometers mentioned at the end of this section).

Small photocell currents may be amplified by conventional electronic techniques, but noise, as well as signal, is amplified. Photomultipliers provide one solution to the problem. A photomultiplier consists of an electron multiplier coupled to a photocathode. Light falling upon this cathode produces electrons via the photoelectric effect; the electrons proceed to hit further active surfaces (*dynodes*), which in turn release further electrons, so that the current produced is amplified.

1.27 Thermopiles can provide intensity in absolute units (watts, or photons s^{-1}), since they depend on a heating effect; the other devices give relative intensities only, and need to be calibrated for absolute measurements.

Absolute calibration procedures using thermopiles are awkward and require great care to avoid errors. It is common practice to employ *chemical actinometers* to provide a secondary standard for intensity calibration. Such actinometers are gas- or liquid-phase systems in which the quantum yield has been accurately determined in previous experiments (ultimately referred back to thermopile measurements). Measurement of the reaction rate thus allows calculation of the light intensity. Chemical actinometers are usually chosen to be insensitive to wavelength and other experimental parameters. See Note 2.37 for two well known actinometer systems.

1.28 Scattered light of the incident radiation can cause problems, especially if luminescence is weak, and may in some cases prevent the study of luminescence at wavelengths near that of the exciting light.

Emission studies

The investigation of phenomena such as fluorescence, phosphorescence and chemiluminescence requires the use of specialized techniques, as the intensities of emitted radiation tend to be far lower than the intensities of the light sources discussed in the previous section, so that the detector must be correspondingly sensitive. Although photographic records of emission spectra can provide intensity data averaged over the exposure period, and information about the spectral distribution of the emission, photomultipliers are far more commonly used as a result of their superior sensitivity and speed of response.

The spectroscopic nature of luminescent radiation may be investigated using a dispersing instrument (e.g. monochromator) together with a photomultiplier to detect radiation. Fluorescence or phosphorescence *excitation spectra* are obtained by monitoring the emission intensity (preferably within a narrow wavelength band) as the wavelength of the exciting radiation is altered (see also pp. 43–45).

1.29 It is possible to calculate the quantum yield of emission processes using measurements of either relative or absolute absorption and emission intensities. Another, often more convenient, approach is to compare the emitted intensities from the sample and from a substance whose emission quantum yield is already accurately known.

Fluorescence and phosphorescence are almost always observed at right angles to the direction of the irradiating beam. If an irradiated system exhibits both fluorescence and phosphorescence, it may be difficult to establish the contribution to emission made by each process. The technique used to differentiate between the two phenomena uses time-resolved intensity analysis, of the kind discussed in the next section, to see if any luminescence persists after the illumination is cut off.

Detection of intermediates

The identification of reactive intermediates involved in photochemical change and the measurement of their concentrations are vital in improving understanding of photochemical processes. The term *reactive intermediates* covers species such as the atoms, radicals, and ions that are the primary fragments of photolysis; the excited states of these fragments; and the excited states, including triplets, of the absorber that are first produced and that participate in fluorescence, phosphorescence, and radiationless transitions (IC and ISC). We first review some techniques that are available for the study of intermediates, and then discuss time-resolved studies.

Optical spectroscopy often provides a sensitive and unambiguous way of identifying an intermediate. When the fine structure [Note 1.30] is resolvable, the spectroscopic detail may enable description of the chemical nature, the structure, and the electronic state of the species.

1.30 The *fine structure* in an electronic spectrum arises from the superposition of vibrational and rotational structures on the electronic spectrum, and may thus provide information about geometry (shape and size) and the type of bonding.

Absorption spectroscopy depends upon the species under investigation having a higher state to which an electric dipole transition is allowed. *Laser absorption* studies using tunable dye lasers (see Section 3.7) as the light source are very successful, and are particularly effective for studying atoms and small radicals. Increased absorption can be achieved using a *multipass absorption* experiment, in which the light beam is made to traverse the sample many times using suitably arranged mirrors; lasers are particularly well suited to this purpose as a result of the small divergence of a laser beam. *Resonance absorption* experiments use light sources spectrally matched to the absorption features being studied, e.g. by using electric discharge lamps that contain the appropriate gases. So long as the spectral linewidths of source and absorber match, the sensitivity of absorption can be high, and the specificity excellent.

Optical emission [Note 1.31] can be used to study excited species, if an optical transition occurs with sufficient intensity to a lower state. *Resonance fluorescence* (RF: see p. 44) can be excited by resonance discharge lamps. *Laser induced fluorescence* (LIF) is similar in concept to RF, but uses a tunable laser to excite the resonance transition; LIF provides more versatility and often higher sensitivity than discharge-lamp excited RF.

1.31 Emission studies tend to have a higher ultimate practical sensitivity than absorption studies, as the only radiation comes from the species under investigation, but they are limited to species that have an excited state that lives long enough for spontaneous emission, and generally work on longer timescales than absorption studies. As a result of the high sensitivity, optical pumping to a higher, emitting, state is frequently used to study species initially in their ground state.

Lasers can be used as pumps in *resonance Raman spectroscopy* (RRS) and *coherent anti-Stokes Raman spectroscopy* (CARS). They are also used to great effect in *photoionization* techniques, which work on the basis that it should be possible to ionize a specific intermediate or excited state using monoenergetic photons of the correct energy. The method is sensitive because ions are only produced when the intermediate is present. Multiphoton ionization (MPI) and *resonance-enhanced* multiphoton ionization (REMPI) by lasers permit the ionization of a wide variety of species without the necessity of using vacuum ultraviolet sources. See Section 2.12 for a discussion of multiphoton processes (and REMPI in particular).

Magnetic resonance methods also find a considerable range of application in the study of photochemical intermediates [Note 1.32].

1.32 Some of the magnetic resonance techniques that have found application include *electron spin resonance* (ESR), *optically detected magnetic resonance* (ODMR), *chemically induced dynamic nuclear spin polarization* (CIDNP), *chemically induced dynamic electron spin polarization* (CIDEP) and *laser magnetic resonance* (LMR). [See also Note 2.16 for CIDNP/CIDEP].

Time-resolved experiments

Time-resolved experiments are used to determine the concentration and chemical nature of reactive intermediates and to provide kinetic information about these intermediates. The stationary concentrations of atoms, radicals, or excited species present in a static system are normally too low for the intermediates to be detected by their absorption spectra. In *flash photolysis*, the archetypal time-resolved technique, an extremely high-intensity flash source is used so that the transient concentrations of intermediates may be sufficiently large for spectroscopic observation. The concentrations of reactants, intermediates, or products can be followed as a function of time after the flash, a process often known as *kinetic spectroscopy*. This type of method is invaluable in the elucidation of photochemical reaction mechanisms.

In a 'pump-and-probe' procedure, a record ('probe') of the absorption spectrum is obtained at a fixed time after the photolysis flash ('pump'). A convenient implementation for kinetic experiments is *flash spectrophotometry*, in which a broad-band continuous-wave source and monochromator combination, or a laser, set at the wavelength of absorption of the species of interest, are used to measure the transmitted intensity as a function of time [Note 1.34].

Luminescent lifetimes may be studied in a manner wholly analogous to absorption flash spectrophotometry, the monitoring light source being omitted (see also p. 61). A further variant is the use of resonance fluorescence (RF) or laser induced fluorescence (LIF) to probe the concentrations of non-emitting intermediates produced by the photolytic flash.

Experiments on the millisecond and microsecond timescale provide information about the rates of bimolecular reaction of photolytic fragments and excited states, as well as about the emission of radiation from triplets in phosphorescence. Nanosecond experiments are able to probe the fluorescent emission from singlets, as well as intersystem crossing, while picosecond studies yield kinetic data about geminate recombination (see Section 2.5), energy exchange, vibrational relaxation, and some of the slower internal conversions and isomerizations. Investigations on the femtosecond timescale have been reported, although it should be borne in mind that in one femtosecond light travels only 300 nm, or one wavelength! Experiments on these timescales probe the act of absorption of radiation and the very first stages of utilization of the energy in promoting chemical and physical change.

1.33 In the earlier experiments, a second flash was used to obtain a photographic record of the absorption spectrum of the reaction mixture at a fixed time after the photolysis flash. The reaction vessel was then filled with fresh reactant, and the experiment repeated with a different delay time between the two flashes.

1.34 The type of method described in the text can be modified to operate on a nanosecond timescale. Experiments on a picosecond or shorter timescale require rather different approaches, which are described in a number of the texts listed in *Further reading* (p. 89).

1.35 The phenomena and processes mentioned in the text will be explained and discussed at later stages of the book.

2 Chemical Change

2.1 Types of chemical reaction

The different possible fates of an excited species were outlined in Section 1.7. In this chapter, we deal with the processes involving excited species that lead to *chemical* change, i.e. reactions in which the reactants and products differ in chemical identity rather than in state of excitation. Chemical change can come about as a result of photodissociation of the absorbing molecule into fragments, direct reaction of the excited species, or spontaneous isomerization of the excited species. Specific examples are shown in Note 1.18, which illustrates the various pathways outlined in the rest of this section.

Photodissociation (route (i) in Fig. 1.5) is the most obvious type of chemical change; it occurs when the species excited by the absorption of radiation has enough, or more than enough, energy to split into fragments. Several mechanisms are recognized for dissociation processes; the three major routes are *optical dissociation*, *predissociation* and *induced predissociation*, and will be discussed in more detail in Section 2.3. The fragmentation of a species into an ion and an electron, i.e. photoionization (shown in Fig. 1.5 as route (viii)), is a special case of photodissociation.

2.1 Photoionization usually requires wavelengths in the vacuum ultraviolet region [Note 1.24] because the ionization energies of most species are relatively large (and generally larger than chemical dissociation energies).

The chemistry of an excited species may differ markedly from that of the ground-state species, and, as discussed in Section 1.4, the differences can arise as a result of both the excess energy carried by the excited species and the particular electronic arrangement of the excited state. The direct reaction of the excited species (route (ii) in Fig. 1.5) may be either intermolecular (including reaction with added reactants, unexcited molecules of the absorbing substance, or, in solution, with the solvent), or intramolecular (including intramolecular reductions, additions, and various types of isomerization).

The excess energy possessed by an excited species may be sufficient to allow internal rearrangement of the species, i.e. the species may undergo spontaneous isomerization (shown as path (iii) in Fig. 1.5).

2.2 Factors determining reactivity

The factors that contribute to the reactivity of an excited species are all interrelated, and include the excess energy possessed by the species (which may help overcome activation barriers), the intrinsic reactivity of the specific electronic arrangement, and the relative efficiencies of the different competing pathways for loss of the particular electronic state.

One of the most important factors in relation to the electronic arrangement concerns the type of orbital (s, p, σ, or π, etc.) and its symmetry. This point is implicit in much of the discussion that follows in this and the next chapter, and becomes explicit in the *correlation rules* for orbital symmetry and spin that are introduced first at the end of this section [Note 2.2].

2.2 In photochemistry, excited species are necessarily involved, and the reaction pathways allowed by the correlation rules are often different from those for the ground-state partners.

There are a number of related issues, all closely linked to the effect of altering the electronic configuration, that explain why excitation from one state to another can lead to such marked differences in chemistry. These include the effects of changes in geometry, dipole moment, redox characteristics (i.e. electron donating and accepting ability), and the related acid–base properties.

Both the sizes and the shapes of molecules may be affected by excitation and the concurrent redistribution of electrons between bonding, non-bonding and anti-bonding orbitals [Note 2.3]. For a particular excited state, the new steric arrangements can then either increase or decrease reactivity.

The dipole moment of a chemical species is affected by the changes in both electron distribution and molecular geometry. As a result of its dependence on these factors, the dipole moment can give an indication of the possible chemical behaviour of excited states. Both increases and decreases in dipole moment may occur [Note 2.4].

Excited species are usually more effective electron donors *and* acceptors than the ground-state species. Excitation in most molecules involves promotion of a (paired) electron from a lower to a higher orbital, so that its removal from the molecule then requires less energy. At the same time, the promotion of an electron leaves behind a low-lying vacancy which will have a strong tendency to accept another electron. Hence, excitation leads to a reduction in the ionization energy and an increase in the electron affinity of the molecule. One consequence is that the redox properties may be very different from those of the ground-state species. In fact, complete electron transfer to or from an excited molecule is quite common, and many photochemical reactions involve radical anions or cations (see, for example, the discussion of photosynthesis in Section 5.3).

As a consequence of the different donor–acceptor properties, acid–base behaviour can also be influenced by electronic excitation. Experimental evidence shows that the excited states of some molecules may be stronger acids than the ground state, while in other species the excited state is a weaker acid [Note 2.5].

Chemical reactions can be efficient only if the reactants and products are connected by a *continuous* surface that describes the potential energy of the system as a function of the several interatomic distances: a reaction occurring 'on' such a surface is said to proceed *adiabatically*, and the reactants and products are said to *correlate* with each other. In other words, the reactants and products must correlate with each other and with the transition state. The occurrence of chemical reactions is restricted by wave-mechanical arguments, summed up as *correlation rules* (see Sections 2.10 for orbital correlation and 3.5 for electron-spin correlation). Electron spin correlation rules are particularly important, and, as far as the quantum number **S** provides a good description of the system, total electron spin is conserved.

2.3 The ground state of an alkene such as ethene is planar. Promotion of one of the electrons from its π orbital to form the (π,π^*) [Note 1.15] excited state leaves only a σ-bond between the carbon atoms, about which rotation can occur; the configuration adopted, with the minimum overlap between π and π^* orbitals, has the two CH_2 groups lying in perpendicular planes.

2.4 In formaldehyde (methanal, HCHO), there is a 30% *decrease* in dipole moment upon excitation to the (n,π^*) state, indicating reduced polarization in the C=O bond. In an aromatic molecule such as 4-nitroaniline, excitation can *increase* the dipole moment by a factor of more than two, indicating extensive charge transfer in the excited state.

2.5 As an example of altered acidity, consider the S_1 state of phenol, which may be up to 10^6 times as acidic as the S_0 state. This change is attributed to the transfer of electrons from the hydroxy oxygen to the aromatic nucleus. Interestingly, the excited triplet, T_1, is not so highly acidic, because of quantum mechanical effects.

2.3 Photodissociation and predissociation

Optical photodissociation

Our discussion of photodissociation in this book, and the examples we use to illustrate our arguments, will generally focus on diatomic molecules. The two-dimensional representation of the relationship between potential energy and internuclear distance for a diatomic species constitutes the familiar potential energy curve, and the actual form of the curve is frequently known for specific electronic states of diatomic molecules. Although the same physical principles apply to the photochemistry of larger molecules, the descriptions are necessarily more complex and less precise — for instance, the potential *curve* is replaced by a multi-dimensional *surface* in order to define the energy as a function of all the different internuclear distances — and will be covered in more depth later in the chapter.

2.6 Note that absorption spectra give information about vibrational spacings in the upper state of the transition. Information about the lower state can be obtained from the emission spectrum, as explained in Section 3.2 (see Note 3.9 and Fig. 3.4).

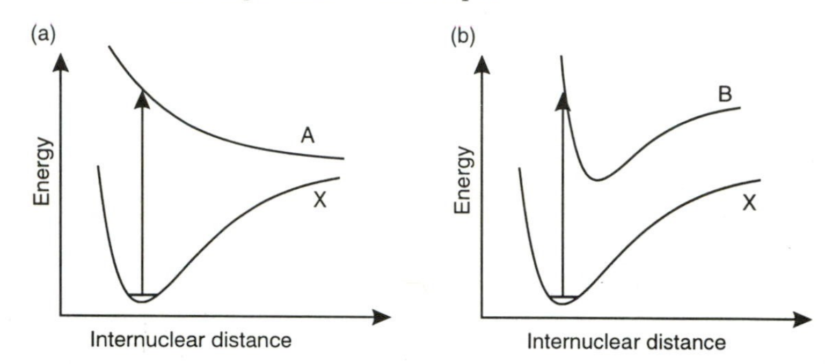

Fig. 2.1 Absorption to (a) an unbound state at an energy greater than its dissociation energy, and (b) a bound state.

2.7 The products of an optical dissociation lie upon the same potential surface as the upper state reached by absorption, i.e. the products correlate with the excited state. It is fairly common in optical dissociation for the upper state to correlate with at least one excited fragment, thus leading to one or more excited product fragments being formed.

Optical dissociation may come about either if absorption occurs so that the excited species possesses energy greater than its own dissociation energy, or if the absorption occurs to an unbound, repulsive state. Both possibilities are shown in Fig. 2.1. In either case, the molecule will dissociate, and any excess energy may be taken up by the fragments as translational energy. Because this translational energy is effectively continuous, the spectrum beyond the point at which dissociation occurs becomes a continuum. If the upper state is bound, then at longer wavelengths (i.e. lower energies) the spectrum may be banded, corresponding to different vibrational levels in the upper state. These levels become progressively closer together, until the continuum is reached: the energy corresponding to the onset of the continuum — the *convergence limit* — is the dissociation energy to the products. Several methods exist to determine the exact convergence limit, but they fall more within the scope of spectroscopy than of photochemistry.

2.8 Photoionization is of profound importance in the upper atmosphere, where short-wavelength ultraviolet radiation from the Sun can lead to appreciable ionization of the chemical species present. One example is the ionization of nitric oxide, which occurs at wavelengths shorter than 134.3 nm

$$NO + h\nu \rightarrow NO^+ + e$$

Complete removal of an electron, *photoionization* [Note 2.8] can also be regarded as a form of photodissociation, yielding dissociation products of an ion and an electron

$$AB + h\nu \rightarrow AB^* \rightarrow AB^+ + e \qquad (2.1)$$

Rydberg series (in which the principal quantum number increases) are known for both atoms and molecules, and the lines or bands converge as the electron is moved into orbitals further away from the nucleus. The convergence limit corresponds to complete removal of the electron, and thus to ionization.

Predissociation

Another route to fragmentation exists; the process involves population of an excited state *below* its dissociation limit, and subsequent radiationless transition to populate another state *above* the dissociation limit of that second state. The phenomenon is referred to as *predissociation* [Note 2.9]. We shall illustrate the process with reference to a model diatomic system, although it is of greater importance in complex organic molecules. Such molecules usually do not undergo direct optical dissociation in the wavelength regions of their strongest absorption, but rather fragment by the predissociation mechanism. Predissociation of these molecules is favoured by the greater number of electronic states, their much closer spacing, and the increased number of vibrational modes, all of which tend to encourage radiationless transitions between different electronic states.

2.9 The gas-phase absorption spectra of the simpler molecules show considerable sharp rotational structure, but in some cases this rotational structure becomes blurred, leading to a diffuseness of the bands at energies well below the optical dissociation limit. The term predissociation was first adopted to describe this spectroscopic situation.

Predissociation in a diatomic molecule can be explained in terms of the 'crossing' potential energy curves of Fig. 2.2. Absorption populates the upper bound curve at an energy above that where the repulsive curve crosses it. Subject to the quantum mechanical requirements for radiationless transition (see Sections 1.9 and 3.4) the system may thus cross to the repulsive state, which falls apart to yield the atomic fragments. In a more complex molecule, the sequence of events is essentially the same, with chemical dissociation occurring from a dissociative state at an energy less than that of the state first populated by absorption. It is important to note that the products may well be different from those yielded by direct optical dissociation [Note 2.10].

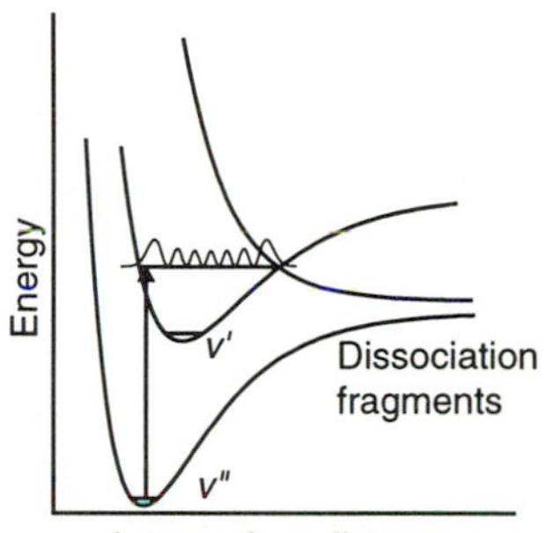

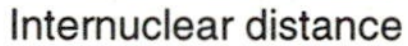

Fig. 2.2 Potential energy curves for a typical diatomic molecule showing the crossing that leads to predissociation.

In certain instances, radiationless transitions that are apparently forbidden by the selection rules can occur in the presence of perturbing influences such as collisions or external electric or magnetic fields (see p. 8). If such radiationless transitions lead to dissociation, the phenomenon is referred to as *induced predissociation*.

2.10 For a diatomic molecule, the fragments from photodissociation must be the same two atoms, whatever the dissociation mechanism. It follows that the state of excitation of fragments produced by predissociation below the optical dissociation limit must be lower than those formed by optical dissociation in a continuous absorption region (see Fig. 2.2).

2.4 Photodissociation: processes and examples

The discussion of photodissociation so far has centred upon diatomic molecules; excluding the case of ionization, the only possible products of dissociation of a diatomic species are the two atomic fragments from which that molecule is formed. The dissociation of polyatomic molecules, however, can yield a multitude of different sets of products, and in order to understand the mechanisms behind such processes it is first necessary to establish the nature of the primary reaction products. In Table 2.1 we provide a selection of the types of process that occur with small inorganic species. Hydrocarbons and carbonyl compounds are then used to illustrate typical photodissociative routes in organic compounds.

Table 2.1 Primary dissociative processes for certain inorganic molecules (from Calvert, J.G., and Pitts, J.N., Jr (1966), Photochemistry).

Species	Products [a]	Wavelength [b] λ(nm)	Quantum Yield [c]
Hydrides			
H_2O	$H + OH(^2\Pi)$	<242	~1
	$H + OH(^2\Sigma)$	<135.6	—
	$H_2 + O(^1D)$	123.6	$\phi(\phi_1+\phi_2) \sim 0.3$
H_2O_2	$2OH(^2\Pi)$	253.7	0.85±0.2
	$OH(^2\Pi) + OH(^2\Sigma)$	<202.5	—
NH_3	$NH_2 + H$	<217	96% of reaction at $\lambda = 184.9$ nm
	$NH(^3\Sigma^-) + 2H$	<155	
	$NH(^1\Pi) + H_2$	<129.5	14% of reaction at $\lambda = 123.6$ nm
	$NH_3^+ + e^-$	<123	
Oxides			
O_2	$O(^3P) + O(^3P)$	~245.4	
	$O(^3P) + O(^1D)$	<175.9	~1
	$O(^3P) + O(^1S)$	<134.2	
O_3	$O(^1D) + O_2(^1\Delta_g)$	<310	~1
	$O(^1D) + O_2(^1\Sigma_g^+)$	<266	—
	$O(^3P) + O_2(^3\Sigma_g^-)$	~600	~1
SO_2	$SO + O$	<218	—
N_2O	$N_2(^1\Sigma_g^+) + O(^1D)$	~180	~1
	$N + NO$		12% at $\lambda = 123.6$ nm
NO_2	$NO + O(^3P)$	<400	
	$NO + O(^1D)$	228.8	
CO_2	$CO(^1\Sigma^+) + O(^3P)$	<227.4	0.06 ($\lambda = 157$ nm)
	$CO(^1\Sigma^+) + O(^1D)$	<167.2	~1
	$CO(^1\Sigma^+) + O(^1S)$	<128.6	($\lambda = 131, 147$ nm)
	$CO(^3\Pi) + O(^3P)$	<108.2	0.75±0.25 ($\lambda = 104.8$ nm)
OClO	$ClO + O(^3P)$	~375.3	—
	$ClO + O(^1D)$	—	
Halogens			
I_2	$I(^2P_{3/2}) + I(^2P_{1/2})$	<499	~1
	$2I(^2P_{3/2})$	<803.7	

(a) The term symbol (see p. 6) is not normally given if a species is formed in its ground electronic state; if, however, both ground state and electronically excited states of the same product can appear, then the states are all specified.

(b) The wavelengths normally refer to those for which a process is energetically possible. If, however, a quantum yield is quoted, the wavelengths refer to those at which the quantum yield was measured or, alternatively, wavelength is specified in the 'quantum yield' column.

(c) Or percentage of total reaction proceeding to the given products.

Hydrocarbons

The alkanes absorb strongly in the vacuum ultraviolet region [Note 1.24]: methane starts to absorb at $\lambda \sim 144$ nm, and the higher alkanes absorb at progressively longer wavelengths; this absorption is thought to be due to an allowed $\sigma \rightarrow \sigma^*$ transition. In the wavelength region 129.5–147 nm, molecular elimination of hydrogen is the most important dissociative process

$$RCH_2R' + h\nu \rightarrow RCR' + H_2 \quad (2.2)$$

although bond ruptures at almost any point are minor processes and can lead to the formation of hydrogen atoms and a variety of free radicals [Note 2.11]. Photoionization occurs at shorter wavelengths (for CH_4 at $\lambda < 96.7$ nm).

2.11 For CH_4, it was hitherto thought that the elimination process was much more efficient than the radical fission

$$CH_4 + h\nu \rightarrow CH_3 + H$$

Recent experiments, however, suggest that at $\lambda = 121.5$nm the radical process dominates.

The lowest-energy singlet → singlet ($\pi \rightarrow \pi^*$) absorption band is found at longer wavelengths in unsaturated hydrocarbons than in the alkanes. Isomerization, which will be dealt with in Section 2.7, is a frequent fate of the excited state formed on absorption, but fragmentation is also observed. For example, in ethene, the processes

$$CH_2{=}CH_2 + h\nu \rightarrow H_2 + H_2C{=}C{:}(\rightarrow HC{\equiv}CH) \quad (2.3)$$
$$\rightarrow 2H + H_2C{=}C{:} \quad (2.4)$$
$$\rightarrow H_2 + HC{\equiv}CH \quad (2.5)$$
$$\rightarrow 2H + HC{\equiv}CH \quad (2.6)$$

are thought to take place over the wavelength range 123.6–184.9 nm. Ethyne is formed in reaction (2.3), via $H_2C{=}C{:}$, about 1.5 times more rapidly than in the direct process (2.5) at $\lambda = 147$ nm. At the same wavelength, the atomic H (2.4), (2.6) and molecular H_2 (2.3), (2.5) eliminations occur with almost equal efficiency; fission to $CH_2CH + H$ is much less common.

Simple aromatic hydrocarbons possess a moderately strong absorption in the near ultraviolet (200–300 nm); absorption in this band leads mainly to emission of radiation or to reaction of the excited species, and quantum yields for photodissociation are very small. However, in the gas phase at $\lambda = 184.9$ nm the quantum yield for benzene disappearance is near unity, and polymer, carbon, and traces of volatile products are formed.

Compounds containing the carbonyl group

Photochemical data are probably more extensive for compounds containing the carbonyl, C=O, group than for any other class of organic compound. In this section, we shall confine our remarks mainly to the photochemistry of aldehydes and ketones, since, broadly speaking, acids, acid anhydrides, esters, and even amides undergo analogous photodissociative reactions.

In each class of carbonyl-containing compound, the first absorption is due to the forbidden $n \rightarrow \pi^*$ transition. In aliphatic aldehydes, the maximum of this band lies near 290 nm, while in ketones it is displaced to slightly shorter wavelengths (~ 280 nm); aromatic substitution shifts the absorption to longer wavelengths (e.g. $\lambda_{max} \sim 340$ nm in benzophenone). The absorption lies at considerably shorter wavelengths in acids, anhydrides and esters (< 250 nm), and in amides (<260 nm). Allowed $\pi \rightarrow \pi^*$ and $n \rightarrow \sigma^*$ transitions give rise to intense absorption bands at short wavelengths (e.g. around 180 and 160 nm for aldehydes).

The important *dissociative* primary photochemical processes in ketones can be represented as

$$\begin{aligned} RCOR' + h\nu &\rightarrow R + (COR')^* \\ &\qquad \hookrightarrow R' + CO \qquad (2.7) \\ &\rightarrow (RCO)^* + R' \\ &\qquad \hookrightarrow R + CO \qquad (2.8) \end{aligned}$$

$$R_2CHCR_2CR_2COR' \rightarrow R_2C{=}CR_2 + CR_2{=}C(OH)R' \\ \hookrightarrow CHR_2COR' \qquad (2.9)$$

The radical-forming processes, (2.7) and (2.8), are known as Norrish Type I photolysis, while the intramolecular fission of the C–C bond α–β to the carbonyl group is known as a Norrish Type II process [Note 2.12].

2.12 The naming of these processes comes from R.G.W. Norrish (Nobel Laureate 1967), who established the existence of the reaction channels in the 1930s.

The relative importance of the two paths (2.7) and (2.8) in Type I photolysis seems to depend on the wavelength of photolysis. Thus, in the photolysis of butan-2-one,

$$\begin{aligned} CH_3COC_2H_5 + h\nu &\rightarrow CH_3CO + C_2H_5 \qquad (2.10) \\ &\rightarrow CH_3 + COC_2H_5 \qquad (2.11) \end{aligned}$$

the relative contribution of paths (2.10) and (2.11) varies from 40:1 at $\lambda = 313$ nm to 2.4:1 at $\lambda = 253.7$ nm.

2.13 Some thermal stabilization of energy-rich acyl radicals may occur in the photolysis of carbonyl compounds, but almost immediate decomposition to alkyl radical and carbon monoxide becomes increasingly likely as the wavelength of photolysis is decreased.

Type II photolysis in ketones has been shown fairly conclusively to proceed via a cyclic six-membered intermediate. For example, in pentan-2-one the process may be represented by the equation

$$CH_3CH_2CH_2COCH_3 + h\nu \rightarrow [\text{cyclic intermediate: } H_2C\text{–}H\cdots O{=}C(CH_3)\text{–}CH_2\text{–}CH_2] \\ \downarrow \\ CH_2{=}CH_2 + CH_2{=}C(OH)CH_3 \qquad (2.12)$$

The *enol* thus formed then tautomerizes to propanone. Further discussion is presented in Section 2.8, since the process involves intramolecular H-abstraction.

Aldehydes can photodissociate according to a Type I mechanism, and, where the structure allows it, a Type II mechanism. An intramolecular elimination reaction, not observed with ketones, also occurs

$$RCHO + h\nu \rightarrow RH + CO \qquad (2.13)$$

Cyclic ketones decompose under the influence of light in a manner which suggests that a diradical is formed in the primary step. Cyclopentanone, for example, is photodissociated in the gas phase in three ways

$$\text{cyclo-}(CH_2)_4C{=}O + h\nu \rightarrow 2C_2H_4 + CO \qquad (2.14)$$

$$\rightarrow \square + CO \qquad (2.15)$$

$$\rightarrow CH_2{=}CHCH_2CH_2CHO \qquad (2.16)$$

The primary products may certainly be interpreted in terms of initial formation of $\cdot CH_2CH_2CH_2CH_2CO\cdot$. It appears that ethene is formed directly from $\cdot CH_2CH_2CH_2CH_2\cdot$ (derived from the primary diradical by loss of CO) rather than from decomposition of 'hot' cyclobutane after ring closure. There is, however, considerable evidence that vibrationally excited molecules are of importance in the gas-phase photolysis.

2.5 Photochemistry in solution

Much of what we have discussed so far has been about gas-phase photochemistry. We now examine the way in which photochemistry may differ in the condensed phase, where the close proximity of neighbouring molecules may have a profound effect on both the process of absorption itself and on the subsequent fate of the species thus excited.

Solvation can reduce molecular energies, to an extent determined both by the nature of the electronic state and by the solvent, so that an isolated species and a species in solution usually absorb at different wavelengths. The energies of ground and excited states may also be reduced to differing extents [Note 2.14].

2.14 Further discussion of the effects of solvents on the energies of ground and excited states is deferred until Section 3.2 (see Note 3.10).

In addition to the observed change in the wavelength of absorption, the *intensity* of absorption may be affected by the change from gas to liquid phase, particularly for forbidden transitions: the disturbances brought about by the neighbouring molecules may be sufficient to invalidate selection rules.

The relative importance of the different fates of excited species may also change as a result of the change from gas phase to solution. For a start, the greatly increased number of collisions undergone by each molecule means that physical quenching becomes far more important, so that the quantum yields for any other primary process are often appreciably smaller than in the gas phase. *Chemical* quenching, i.e. the removal of the excited molecule by chemical reaction, may also become competitive with other loss processes. In addition, the fragments formed by dissociation may react with solvent molecules almost immediately, and the *overall* step may then be regarded for practical purposes as the primary one. This phenomenon is particularly important when radical fragments are formed in a hydrogen-containing solvent, since the formation of *free* radicals is effectively prevented [Note 2.15].

2.15 The influence of H-abstraction from the solvent is exemplified by the photolysis of $(CH_3)_2CH.CO.CH(CH_3)_2$, which in solution yields CO, C_3H_8, and $(CH_3)_2CHCHO$ as the main Type I products. C_6H_{14} and $(CH_3)_2CHCO.COCH(CH_3)_2$ are seen in the gas-phase photolysis but not in solution. The primary C_3H_7 and $(CH_3)_2CHCO$ fragments yield products only by hydrogen atom abstraction.

One factor which is not apparent in the gas phase, but which strongly affects the types and rates of processes taking place in solution, is the occurrence of *cage effects*. In solution, molecules are effectively trapped within a 'cage' of solvent molecules. The separation of reacting species by normal diffusive processes is therefore hindered, and the species may make many mutual collisions before they are eventually liberated from the cage. Cage effects may be particularly important in the *geminate recombination* (from gemini, the Latin word for twins) of radicals formed by the fragmentation of an excited species, since the prevention of their separation encourages very rapid reaction to reform the original product; any excess energy in the newly-formed bond is removed by collision with solvent molecules. Relatively few radicals manage to escape the solvent cage before recombination, and the effective quantum yield for dissociation is consequently substantially reduced.

2.16 Cage effects are responsible for the interesting phenomenon of chemically induced polarization observed in photochemical magnetic resonance experiments. Chemically induced dynamic nuclear spin polarization (CIDNP) manifests itself by abnormal absorption intensities, and even emissions, in NMR studies of photochemical processes. The CIDNP technique affords one method for the diagnostic study of certain types of photochemical reaction. Chemically induced dynamic electron spin polarization (CIDEP) may lead to emission of electron spin resonance (ESR), and the phenomenon can yield information about the energetics and relaxation of triplet states.

The importance of geminate recombination is related to the kinetic energy of the fragments formed on photolysis, as well as to the viscosity of the solvent. If the fragments have sufficient energy, they may force their way out of the cage, and the primary quantum yields of some solution-phase photolyses increase as the wavelength of photolysing radiation is decreased.

2.6 Direct and photosensitized reactions

Intermolecular energy exchange from one species to another, leading to the excitation of a species other than the one absorbing, can be used to promote chemical change that may not be possible through direct absorption (for instance, if the experiment is carried out in a wavelength region where the reactant is transparent). Energy transfer itself is discussed in Chapter 3; we confine our discussion here to the chemical consequences of this process of *photosensitization*. The different route to excitation may lead to the population of excited states that are inaccessible by direct absorption, and unsensitized and sensitized photochemistry may be quite different.

2.17 One example of a photosensitized reaction involves methane: CH_4 is virtually transparent at $\lambda > 170$ nm; however, in the presence of mercury vapour, irradiation with the mercury resonance line at $\lambda = 253.7$ nm dissociates CH_4. The sequence of events can be written

$$Hg + h\nu_{\lambda=253.7\ nm} \rightarrow Hg^*$$
$$Hg^* + CH_4 \rightarrow CH_3 + H + Hg$$

Excited singlet states and triplet states of organic compounds may react in distinct ways. For example, the direct photoisomerization of *trans*-penta-1,3-diene (2.17), which involves the rearrangement of an excited singlet reactant, gives different products from those of the sensitized reaction (2.18), which involves the triplet

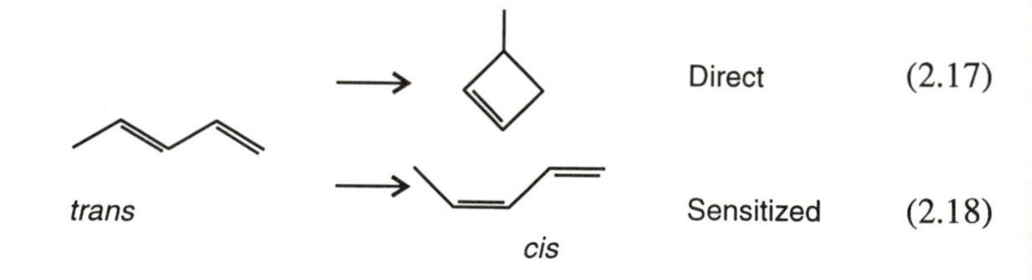

Another similar example will be encountered in the next section (reactions (2.21) and (2.22)).

Many different substances have been used as sensitizers. The energy transfer step is assumed to be adiabatic, and hence spin conserved (see p. 50), with organic sensitizers, and the occurrence of photosensitized reactions provides valuable information about intersystem crossing and triplet chemistry. Ketones have proved very popular for triplet sensitization studies, because triplet yields in carbonyl compounds are high, and the triplets of many acceptors lie lower in energy than carbonyl triplets and can therefore be populated efficiently in an exothermic energy transfer process.

2.18 Benzophenone is frequently used in condensed-phase experiments employing sensitization, while biacetyl is particularly useful for gas-phase work. Historically, mercury was popular for gas-phase studies because of its volatility at room temperature, and the ease with which emission lamps for the $\lambda = 253.7$ nm resonance line can be constructed.

2.7 Isomerizations and rearrangements

A wide variety of isomerizations and rearrangements can be induced photochemically, and in this section we shall examine some of the ways in which isomerization can be brought about by the absorption of light.

The lowest excited singlet and triplet states of ethene and its derivatives are (π, π^*) in character [Note 1.15], formed by promotion of an electron from the highest filled bonding π orbital to the lowest unfilled antibonding π^* orbital. This change in electronic configuration removes the constraints that hold the molecule in a fixed planar geometry, so that free rotation can then occur about the carbon–carbon axis [Note 2.3]. Both S_1 and T_1 states adopt a perpendicular configuration, so that excitation of both *cis* and *trans* isomers gives rise to a geometrically equivalent excited state. The subsequent relaxation to the ground state is accompanied by a necessary re-adoption of the planar configuration, with both *cis* and *trans* isomers being formed.

2.19 Geometrical photoisomerization can occur from *cis* to *trans* or from *trans* to *cis* isomers, so that on prolonged irradiation of either isomer, a stationary state is set up. If the quantum yields for processes in each direction are similar, the concentrations of *cis* and *trans* isomers at the stationary state are determined mainly by the amount of light absorbed (and thus by the extinction coefficient) of each isomer at the wavelength employed. For many simple alkenic systems, the higher extinction coefficient at longer wavelengths is possessed by the *trans* isomer, and if long wavelength light is used, the *cis* isomer will predominate at the photostationary state.

Photochemical valence and structural isomerizations are well known. The electrocyclic ring-closing reactions of dienes and trienes are typical of the valence isomerizations (see Section 2.10), and are typified in the formation of cyclobutene from butadiene

(2.19)

These electrocyclic processes require the *cis* conformation. An alternative cyclization yields a bicyclobutane

(2.20)
[Note 2.20]

The reactions of polyenes are often different for excited singlet and triplet states. Triplet-sensitized reaction often leads to (cyclic) dimerization (see Section 2.10), while the direct reaction leads to internal cyclization.

2.20 Since bicyclobutane is also formed in the triplet- sensitized reaction, it is possible that ring closure involves a diradical in a non-concerted process.

Entirely different products are observed on direct and sensitized irradiation of myrcene

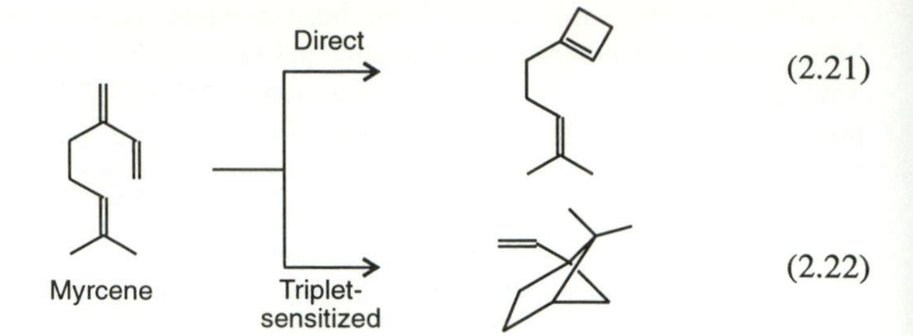

(2.21)

(2.22)

Since the products of reaction (2.22) are probably formed via the triplet of the triene, it may be inferred that the singlet is responsible for the products of the direct photolysis.

Aromatic compounds often cyclize to give polycyclic, non-aromatic products, as exemplified by the photocyclization of *cis*-stilbene to dihydrophenanthrene

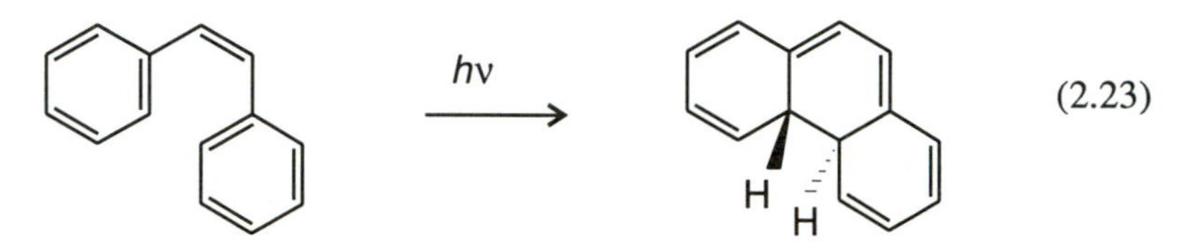

(2.23)

a process that accompanies direct *cis*- and *trans*-isomerization. Valence isomerization of benzene itself produces the highly reactive species benzvalene (and, at short wavelengths, bicyclohexadiene and fulvene as well)

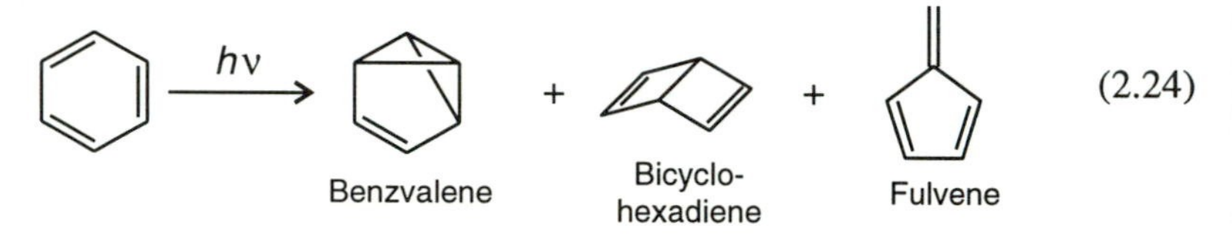

(2.24)

Another important photoisomerization, in which 1,4-dienes and 3-phenylalkene systems yield vinylcyclopropanes, is known as the *di-π*-methane rearrangement [Note 2.21]

2.21 The mechanism for the *di-π*-methane rearrangement is formally a 1,2-shift with ring closure.

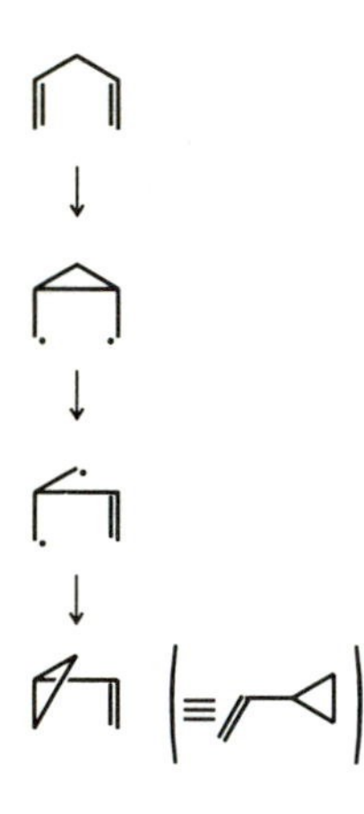

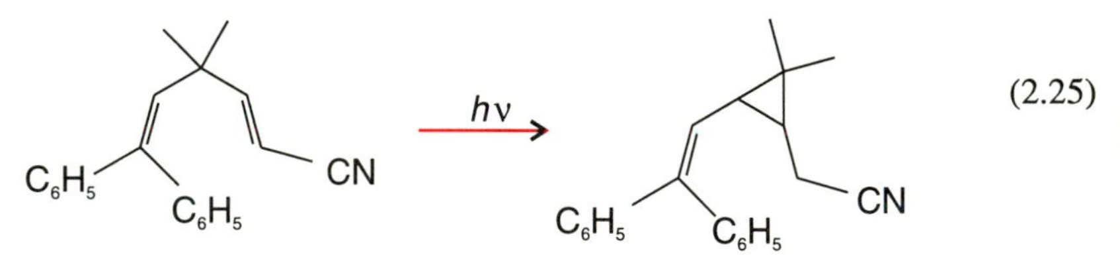

(2.25)

2.8 Intramolecular abstraction reactions

One of the most important of all the various possible intramolecular reactions undergone by excited species, and certainly the most common of intramolecular abstractions, is the abstraction of hydrogen. It occurs in species with lowest excited states that are (n, π*) in character, notably carbonyl compounds. Those 'abnormal' carbonyl compounds whose lowest excited levels are (π,π^*) states [Note 2.22] undergo neither intramolecular nor intermolecular H-abstraction, except with very good donors such as amines that provide hydrogen by an indirect route. It is interesting to note that this hydrogen abstraction from carbonyl compounds is part of the sequence of events leading to the Norrish Type II fragmentation (see Section 2.4). Equation (2.12) shows a six-membered transition state in the Type II fission of a ketone: this cyclic intermediate favours intramolecular H-abstraction over intermolecular abstraction from the solvent, and the final steps from the cyclic intermediate in reaction (2.12) may be written as

$$CH_3CH_2CH_2COCH_3 + h\nu \longrightarrow \text{(cyclic: } H_2C(H)\cdots O{=}C(CH_3)\text{, } H_2C{-}CH_2) \xrightarrow{\text{intramolecular H-abstraction}} H_2\dot{C}{-}CH_2{-}CH_2{-}\dot{C}(OH){-}CH_3 \longrightarrow CH_2{=}CH_2 + CH_2{=}C(OH)CH_3 \longrightarrow CH_3COCH_3 \qquad (2.26)$$

In addition to the fragmentation reaction of the hydroxy diradical formed by intramolecular H-abstraction, there is a ring-closure path leading to cyclobutanol formation

$$R{-}\dot{C}(OH){-}CH_2CH_2\dot{C}H_2 \longrightarrow R{-}C(OH)\text{(cyclobutane: } {-}CH_2{-}CH_2{-}CH_2{-}) \qquad (2.27)$$

In certain cases, the photochemical cyclobutanol formation appears to be stereospecific, e.g.

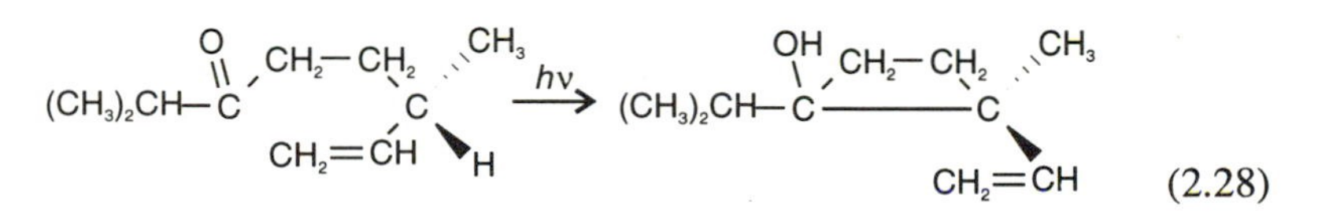

(2.28)

2.9 Intermolecular abstraction reactions

The abstraction of hydrogen by reduction of excited (n,π^*) states, especially those of aromatic carbonyl compounds, is not necessarily confined to intramolecular processes, but may also occur via *inter*molecular interactions. A typical example is the photoreduction of benzophenone in the presence of a suitable hydrogen donor

$$C_6H_5COC_6H_5 + RH \xrightarrow{h\nu} C_6H_5\dot{C}(OH)C_6H_5 + \dot{R} \qquad (2.29)$$

If the solvent is a good hydrogen donor, such as ethanol, then the quantum yield for the disappearance of benzophenone may be almost one. If the solvent is a 'reluctant' donor, and does not willingly give up hydrogen, then the quantum yield can be much smaller. If propan-2-ol is used as a solvent, then

2.22 In 'normal' carbonyl compounds such as benzophenone, the lowest triplet is (n,π^*) in character. For some 4-substituted ketones (e.g. 4-aminobenzophenone) the lowest triplet is a (π,π^*) state. The reactivities of normal and abnormal compounds are entirely different, which is hardly surprising given that different types of electron belonging to different parts of the molecule have been promoted.

2.23 Diradicals have been detected in the Type II fission of a ketone (e.g. by flash spectroscopy), and the enols produced by cleavage have also been investigated spectroscopically at low temperatures.

2.24 Higher yields of the cyclization product can also be obtained by using ordered arrays of molecules (e.g. in micelles). It seems likely that the orbitals of the diradical electrons and of the bond to be broken must be aligned parallel for efficient cleavage to occur.

2.25 The 'abnormal' ketones [Note 2.22] do not participate efficiently in the abstraction process.

under suitable conditions quantum yields around *two* can be obtained for benzophenone removal. The radical produced from the solvent is itself capable of reducing a molecule of benzophenone to its ketyl radical, and two molecules of benzophenone are removed for each quantum of light absorbed

$$(C_6H_5)_2CO + (CH_3)_2CHOH \xrightarrow{h\nu} (C_6H_5)_2COH + (CH_3)_2COH \qquad (2.30)$$

$$(CH_3)_2COH + (C_6H_5)_2CO \rightarrow (CH_3)_2C{=}O + (C_6H_5)_2COH \qquad (2.31)$$

2.26 The ketyl radicals formed in reactions (2.30) and (2.31) can participate in secondary reactions, which include dimerization to form a pinacol

$$C_6H_5-\underset{C_6H_5}{\overset{OH}{C}}-\underset{C_6H_5}{\overset{OH}{C}}-C_6H_5$$

or further hydrogen abstraction (in alkaline solution) to form benzohydrol, $(C_6H_5)_2CHOH$. Photopinacolization has been recognized since the beginning of the century, when it was found that good yields of benzpinacol were produced by the action of sunlight on solutions of benzophenone.

The efficiency of photoreduction depends not only on the nature of the solvent, but also on the structure of the ketone. Substitution of aryl ketones in the 2-position makes possible the *intra*molecular hydrogen abstraction discussed in the last section. A six-membered cyclic transition state is available, and photopinacolization [Note 2.26] is replaced by the formation of an unsaturated compound [Note 2.27]

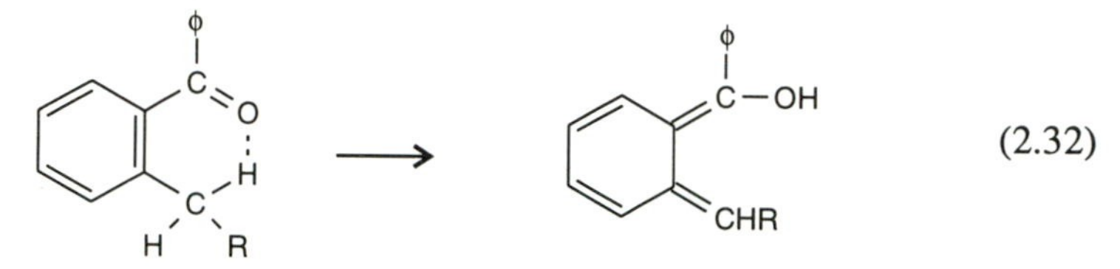

(2.32)

2.27 Steric hindrance is *not* the main factor in reducing the efficiency of intermolecular hydrogen abstraction, because where the 2-substituent has no available hydrogens (i.e. where the six-membered ring cannot be formed), pinacols are formed quite efficiently.

2.10 Photoaddition and photocyclization reactions

A multitude of photochemical addition reactions can occur, both homoadditions and heteroadditions being known. Alkenes can undergo photochemical electrophilic addition, for example with water, alcohols, and carboxylic acids. The route of many of these reactions appears to be determined by quantum mechanical 'correlation' rules (see Section 2.2 and later in this section) that apply to the orbital symmetry. Photoadditions involving the excitation of an aromatic compound are also possible, for instance in the reaction with an amine

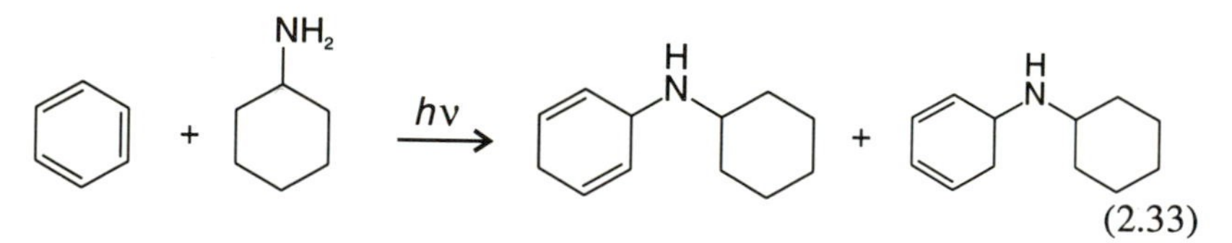

(2.33)

Some of the most interesting photochemical addition reactions are photocyclizations, in which two new σ-bonds are formed to yield a cyclic product. Common classes of cycloaddition include the (2 + 2) addition of two alkenes to give a cyclobutane

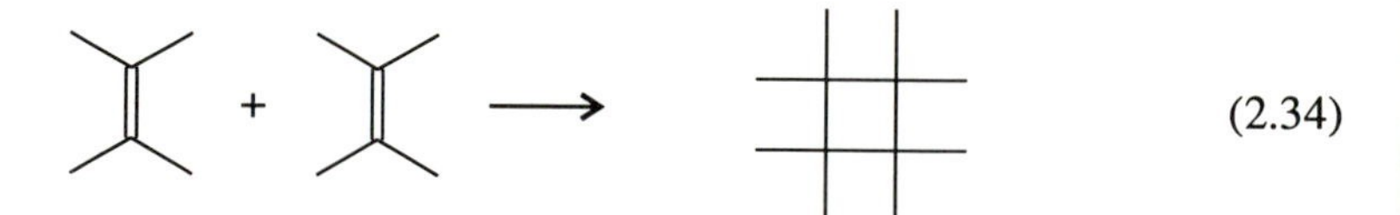

(2.34)

and the (4 + 2) addition of a conjugated diene to an alkane to give a cyclohexene

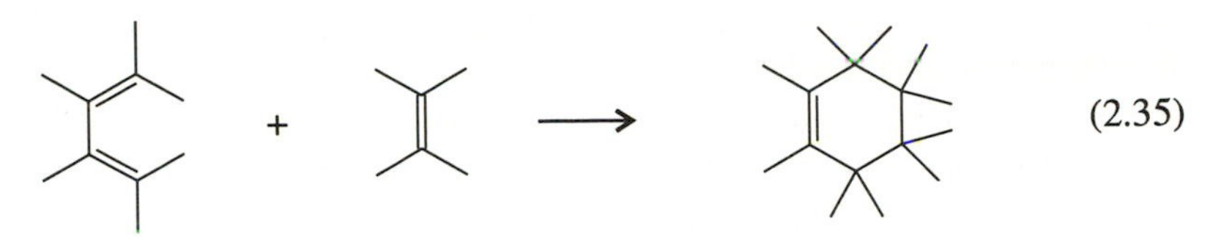

The majority of the cycloaddition reactions of alkenes occur via a two-step diradical mechanism. In some cases, however, (2+2) cycloadditions may occur via a single, *concerted*, step. This type of reaction is allowed on orbital symmetry grounds so long as it occurs in a suprafacial–suprafacial manner [Note 2.28]. The dimerization of but-2-ene provides an excellent example of such a concerted reaction. Irradiation of either the *cis* or the *trans* isomer of but-2-ene on its own gives rise to a pair of isomers of the product, 1,2,3,4-tetramethylcyclobutane, with the stereochemistry expected for concerted addition. Triplet sensitizers appear to promote the more usual diradical mechanism.

2.28 *Suprafacial* refers to addition to the same side of a π-system. Addition to opposite sides is denoted *antarafacial.*

Benzene and its derivatives can undergo cycloaddition across the 1,2-, 1,3-, or 1,4-positions. The 1,3-addition is the main reaction with simple alkenes

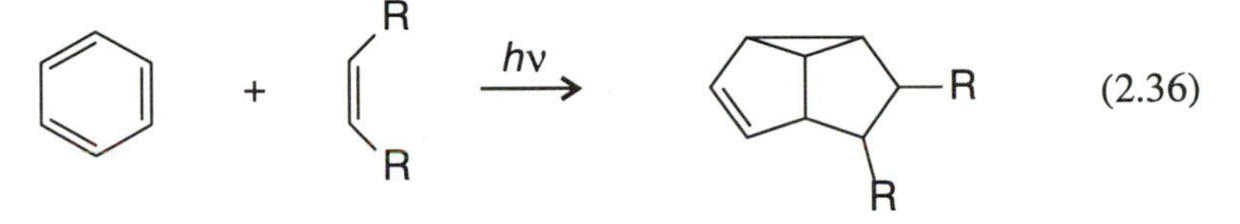

but it seems to be restricted to double bonds bearing only alkyl substituents. Addition is stereospecific, and involves singlet excited benzene.

2.29 Open-chain dienes, such as buta-1,3-diene, form cyclobutane derivatives

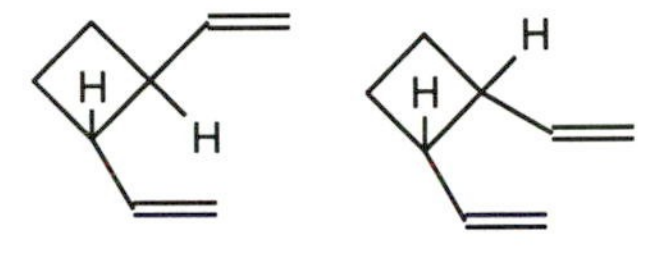

and, in addition, 4-vinylcyclohexene.

Cyclobutane derivatives are formed by the photoaddition of alkenes to the unsaturated bond in α,β-unsaturated ketones. For example, cyclobutane derivatives are the major products of irradiation of cyclohex-2-enone in the presence of 2-methylpropene

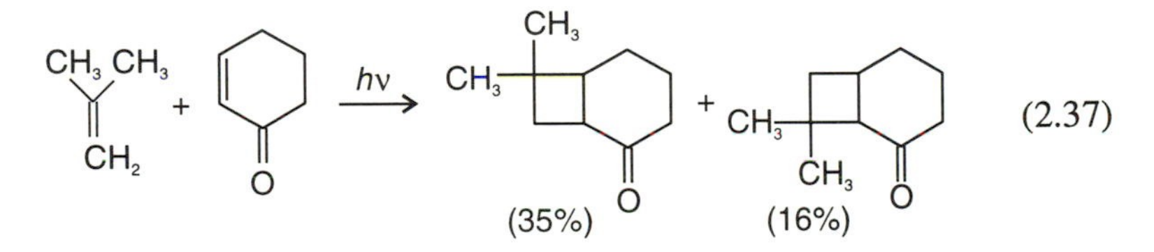

Excited aldehydes and ketones may participate in reactions involving the addition of the excited carbonyl group itself to suitable alkenes, to give an oxetane. For example, irradiation of benzophenone in the presence of 2-methylpropene leads to the formation of an oxetane in relatively high yield

$C_6H_5CC_6H_5$ (C=O) + $(CH_3)_2C{=}CH_2$ $\xrightarrow{h\nu}$ oxetane: CH_3, C_6H_5, CH_3, C_6H_5 (2.38)

The triplet (n, π*) state of the carbonyl compound appears to attack the ground state of the alkene to yield a 1,4-diradical, which then cyclizes.

Photosensitized oxidation

Many of the photosensitized oxidations undergone by unsaturated compounds are thought to involve the photochemical addition of oxygen. Photosensitized oxidations are of particular importance in the biological field, because of their impact on living organisms. It has long been known that microorganisms can be killed in the presence of oxygen and a sensitizing dye. The pathological effects of photooxidation of cell constituents are now known to include cell damage, induction of mutations or cancer, and death. Recent investigations of photosensitized oxidation have led to a better understanding of the chemical processes, and the results are now finding application in the biological field. Some brief remarks about photomedicine are presented in Section 6.10.

The majority of photosensitized oxidations involve the sensitizer in its triplet state, since the singlet is relatively so short lived. The sensitizer triplet may then react either with the unsaturated compound itself, or with the oxygen. For many sensitizers, the reaction with oxygen is so efficient that, at all but the lowest oxygen concentrations, the latter process is favoured over the former. The chemical nature of the unsaturated compound determines whether the presence of oxygen merely inhibits the former possibility, or whether the products of the primary reaction with oxygen eventually lead to oxidation of the substrate. The sensitized oxidations of alkenes, dienes, dienoid heterocycles, and polycyclic aromatic compounds are particularly efficient, and it is with these substances as oxidizable reactant that we are concerned. The first oxidation products are often peroxides or hydroperoxides, and they may subsequently take part in secondary oxidation steps.

Overwhelming evidence points to a mechanism for these photooxidations in which energy transfer from the (triplet) sensitizer to (triplet) ground state oxygen occurs

$$^3\mathrm{Sens} + O_2 \rightarrow \mathrm{Sens} + O_2^* \qquad (2.39)$$

Since the ground state of oxygen is a triplet, O_2^* must be a singlet state for spin to be conserved in reaction (2.39) (see p. 50 for an explanation). The lowest-lying excited state of oxygen is, in fact, a singlet ($^1\Delta_g$) [Note 2.30], and possesses an energy of 92 kJ mol^{-1}, so that it can easily be excited by energy transfer from the triplet states of most dyes.

Direct studies of the reactions of $O_2(^1\Delta_g)$ have indicated that it is the intermediate involved in sensitized photooxidation. It can be produced in a number of ways: for example, in the reaction of sodium hypochlorite with hydrogen peroxide, or by the action of a microwave discharge on molecular oxygen in the gas phase. For a wide variety of acceptors that yield more than one oxidation product, the product distributions from the reaction with $O_2(^1\Delta_g)$ and from the photooxidation are identical, and there are no detectable differences in stereoselectivity.

As typical examples, we may consider the products formed on irradiation of dimethylfuran and of tetramethylethene in the presence of oxygen and a triplet sensitizer. In the first case, an endoperoxide is formed, while the

2.30 The state of O_2 is given as $^1\Delta_g$, which is a *term symbol* (see p. 6) whose explanation can be found in any text on spectroscopy. For the present purposes, it should be considered merely as a label to identify O_2^*.

2.31 The elucidation of the mechanism of sensitized photooxidation has made possible several fruitful speculations with regard to photobiology. An example is the protective action of carotenoids in biological systems. Carotenoids apparently protect photosynthetic organisms against the lethal effects of their own chlorophyll, which is an excellent sensitizer of photooxidation. It has been shown that β-carotene is an extremely efficient quencher of singlet oxygen, and it can also inhibit sensitized photooxidations. Studies suggest that the quenching of $O_2(^1\Delta_g)$ may involve excitation of triplet carotene by energy transfer. Thus, the interesting speculation may be made that carotenoids serve a double function in photosynthetic organisms (see Section 5.3): first, that they remove 'toxic' singlet oxygen, and, secondly, that they can store the energy that O_2 receives from chlorophyll, which would otherwise be lost.

product of the alkene oxidation is a hydroperoxide

$$H_3C\text{-(furan)-}CH_3 + O_2 \xrightarrow{h\nu,\ \text{sens}} H_3C\text{-(endoperoxide)-}CH_3 \qquad (2.40)$$

$$(H_3C)_2C{=}C(CH_3)_2 + O_2 \xrightarrow{h\nu,\ \text{sens}} (H_3C)_2C(OOH){-}C(CH_3){=}CH_2 \qquad (2.41)$$

Alkenes with electron-rich double bonds, or those that do not possess an allylic hydrogen, can undergo a (2+2) cycloaddition to form a dioxetane

$$C_2H_5O{-}CH{=}CH{-}OC_2H_5 + O_2 \xrightarrow{h\nu,\ \text{sens}} C_2H_5OCH{-}CHOC_2H_5\ (\text{O{-}O bridged}) \qquad (2.42)$$

Dioxetanes can be quite stable, but they undergo chemiluminescent (see Section 3.8) thermal decomposition to produce two carbonyl fragments, one of which is electronically excited.

Correlation rules

Photocyclization reactions such as reaction (2.19) and photoadditions such as reaction (2.37) often occur in a manner that is controlled by quantum-mechanical *correlation rules* (see also Sections 2.2 and 3.5). *Pericyclic* reactions are characterized by their concerted nature and the participation of a cyclic transition state. The three main classes of pericyclic reaction are *electrocyclic* reactions, which involve ring closure in conjugated π-systems or its reverse, *sigmatropic* reactions, in which a σ-bond migrates with respect to a π-framework, and *cycloadditions* and their reverse.

2.32 As used frequently in this book, the word 'correlation' means that the reactants and products are connected by the same potential energy surface (see p.19), and that crossing from one surface to another does not take place. A reaction of this kind is termed *adiabatic* and its occurrence implies the connection of spin, orbital angular momentum, and certain symmetry properties. It is the conservation of symmetry that lies behind the Woodward–Hoffman rules. Spin conservation, as discussed in Section 3.5, is another consequence of quantum mechanical correlation rules.

Some of the most thorough and innovative work in the field of pericyclic reactions was carried out by R. B. Woodward and R. Hoffman. They developed a system in which they used the ideas of orbital symmetry to predict not only the types of cyclic transition state that were energetically permitted, but also the stereochemical consequences of the reaction. Their results are summarized in the *Woodward–Hoffman rules*.

To illustrate the method, we shall use as our example the cyclization of a substituted buta-1,3-diene to a form cyclobutene (cf. reaction 2.19)

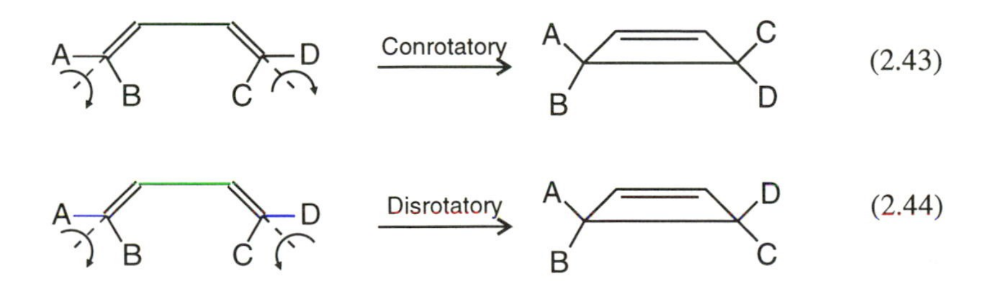

The ring is effectively closed as a result of formation of a σ-bond by the overlap of p-orbitals on C1 and C4 of the butadiene. There are two different possibilities of geometrical change that can bring about this overlap: *conrotatory*, as in reaction (2.43), and *disrotatory*, as in reaction (2.44).

To find out whether conrotatory or disrotatory ring closures are allowed, it is necessary to discover which electronic states of the reactant correlate [Note 2.32] with which of the product. For the conrotatory electrocyclization of the substituted butadiene, it can be shown that S_0 of the reactant correlates with S_0 of the product, and the reaction can therefore proceed adiabatically from the ground-state reactant. For disrotatory electrocyclization, the S_0 states do not correlate and the reaction therefore cannot proceed adiabatically. The conclusion is that *thermal* reaction is allowed for conrotatory cyclization, but is forbidden in the disrotatory mode.

In a similar way, it can be demonstrated that the S_1 state of butadiene correlates with S_1 in cyclobutene for disrotatory ring closure, thereby implying that the disrotatory reaction might be allowed in photochemically induced cyclization. Evidence corroborating this assumption has been provided by experiments that successfully demonstrate the occurrence of photochemical disrotatory ring closure.

It has been found experimentally that, in general, the pericyclic reactions of ground and excited states proceed in opposing geometrical modes. For example, thermal cyclodimerization of substituted ethenes occurs predominantly by *cis–trans* addition, while photocyclodimerization yields mostly *cis–cis* addition products. Correlation diagrams provide an effective interpretation of the contrasting courses of thermal and photochemical cyclizations.

2.11 Ionic species, and charge and electron transfer

In Section 2.5, we discussed some of the special aspects of solution-phase photochemistry. One of the most important features, especially in polar solvents (and particularly water), is the significance of ionic species. 'Primary' photochemical reactions involving ions in solution may be very different from the photodissociation of neutral molecules, since oxidation–reduction steps are frequently involved.

The characteristic colours of the complexes of transition metal ions usually arise from optical transitions of electrons on the metal atom. These absorptions are frequently fairly weak, as the transitions are forbidden on symmetry grounds. Many ions do, however, show a strong absorption in the ultraviolet region (usually between 200 and 250 nm), which is due to *charge-transfer* transitions, i.e. transitions in which an electron is transferred from one ion to another, or from an ion to the solvent [Notes 2.33 and 2.34].

The reducing species formed on photolysis of aqueous solutions of ions appears to be a hydrated electron. Much of the evidence from which this

2.33 The dark colour of the complex halides of Cu^{2+}, as distinct from the light blue of hydrated Cu^{2+}, is due to the tail of a charge-transfer spectrum (most of which lies in the ultraviolet).

2.34 Electron-transfer processes are of great importance throughout condensed-phase chemistry. The efficiency of such processes sometimes shows a dependence on the free energy of reaction that is somewhat surprising, with the rate constant *decreasing* with increasing free energy in the so-called 'inverted' region. This behaviour is the subject of *Marcus theory*, which explains the anomalies in terms of the wave-like properties associated with electrons. It is thought likely that some of the steps in photosynthesis (see Section 5.3) and some other natural phenomena can occur only because of these special features of photoinduced electron transfer.

conclusion has been drawn comes from comparisons with the products of the radiolysis of water (i.e. a process in which water is stripped of an electron). For instance, a transient absorption at around 700 nm is observed on flash photolysis (see Section 1.10) of aqueous ionic solutions, which appears identical to the absorption seen on pulse radiolysis of pure water. In addition, the rates of reaction of the species produced in the two ways are also often identical. Furthermore, both optical and electron paramagnetic resonance spectra of uv-irradiated aqueous glasses containing ionic species, and of trapped electrons produced by ionizing radiation, are identical. The wavelengths involved imply that much less energy is required to remove the electron from the parent species than in gas-phase photoionization: this energy difference presumably arises as a result of the stabilizing effect of hydration of the electron. Quantum yields for the formation of hydrated electrons have been estimated, and may be relatively high. In photolysis of the halide ions (I^-, $\lambda = 253.7$ nm; Br^-, Cl^-, $\lambda = 184.9$ nm), for example, the quantum yields are probably 0.3–0.5.

2.35 Although equations showing transfer of charge are written as if an electron is completely transferred, it must be understood that they may also represent partial electron transfer.

If the hydrated electron is involved, then the photochemical process may be written

$$X^{n+} + H_2O + h\nu \rightarrow X^{(n+1)+} + H_2O^- \qquad (2.45)$$

The hydrated electron may then spontaneously dissociate,

$$H_2O^- \rightleftharpoons OH^- + H \qquad (2.46)$$

or, more particularly, in acid solution it may react according to

$$H_2O^- + H^+ \rightarrow H_2O + H \qquad (2.47)$$

2.36 The reactions of H_2O^+, analogous to (2.46) and (2.47), are suggested to be

$$H_2O^+ \rightleftharpoons OH + H^+$$
$$H_2O^+ + OH^- \rightarrow OH + H_2O$$

In either case, atomic hydrogen is produced, and it can initiate secondary radical reactions. The photochemical decomposition of the I^- ion, for instance, shows a pH dependence that is consistent with a chain mechanism initiated by the sequence of steps

$$I^- + H_2O + h\nu \rightarrow I + H_2O^- \rightarrow I + H + OH^- \qquad (2.48)$$

and the low *overall* efficiency can be attributed to the recombination of H and I atoms within the solvent cage (see Section 2.5).

Charge transfer in cations can occur either to or from the solvent. Iron-containing solutions can therefore show absorption bands for both types of charge transfer

$$Fe^{2+}{\cdot}H_2O + h\nu \rightarrow Fe^{3+}{\cdot}H_2O^-, \quad \lambda_{max} \sim 285\text{ nm} \qquad (2.49)$$
$$Fe^{3+}{\cdot}H_2O + h\nu \rightarrow Fe^{2+}{\cdot}H_2O^+, \quad \lambda_{max} \sim 230\text{ nm} \qquad (2.50)$$

The longest wavelength absorption appears to correspond to the direction of the oxidation–reduction reaction that proceeds most easily.

2.37 Charge transfer occurs in two substances, $UO_2C_2O_4$ and $K_3Fe(C_2O_4)_3$, that are frequently used as *chemical actinometers*. In $UO_2C_2O_4$, charge transfer from UO_2^{2+} to $C_2O_4^{2-}$ ions leads to decomposition of the $C_2O_4^{2-}$; in $K_3Fe(C_2O_4)_3$, the change of importance is the reduction of Fe^{3+} ions to Fe^{2+}.

2.12 Multiphoton photochemistry

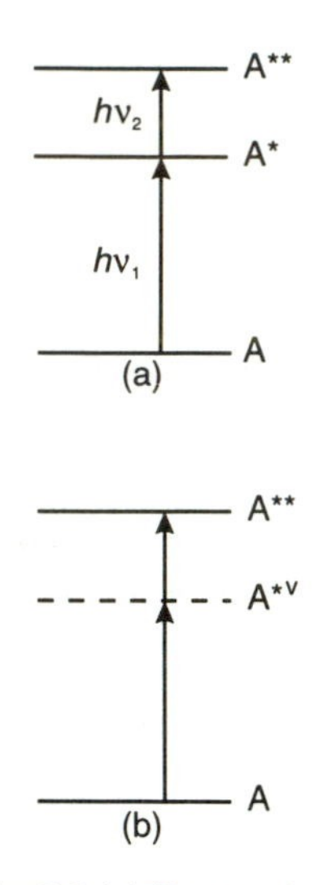

Fig. 2.3 (a) Resonant and (b) non-resonant two-photon absorption.

A relatively recent development in the study of photochemical reactions involves *multiphoton* processes, in which a single particle absorbs more than one photon. Such absorption only occurs at sufficiently high intensities, and it has taken the development of lasers (see Sections 3.7 and 6.8) to provide a suitable light source with which the study of such reactions is feasible.

At first appearances, multiphoton processes seem to be in contradiction to the basic tenets of photochemistry as embodied in the Stark–Einstein laws and as explained in Section 1.8. Despite the apparent inconsistency, however, multiphoton processes do not violate quantum theory; the simple view must merely be modified to allow for the situation that arises when more than one photon can arrive at a single particle in very rapid succession.

Our discussion of multiphoton processes will concentrate on two-photon excitation, and the ideas presented can readily be extended to account for a greater number of photons. Two mechanisms may be considered in the achievement of multiphoton excitation. *Sequential* excitation involves a real intermediate state of the absorbing species. That state becomes populated by the first photon, and it can act as the starting point for the absorption of the second, as illustrated in Fig. 2.3(a). The real intermediate state A* has a well-defined lifetime, typically 10^{-4} to 10^{-9} s. Because it is excited by a resonantly-absorbed photon, the overall sequential process is referred to as *resonant two-photon excitation.* The second excitation mechanism is *non-resonant two-photon excitation*, in which no resonant intermediate states are involved [Fig. 2.3(b)]. A *virtual* state, A^{*v}, is created by the interaction of the first photon with A. Only if a second photon arrives within the duration of the first interaction (about 1 cycle of the electromagnetic radiation, or 10^{-15} s) can it be absorbed. The process is therefore often referred to as *simultaneous* biphotonic absorption, to distinguish it from the sequential mechanism. It is now apparent why high intensities are essential to achieve two-photon excitation. In the resonant process, it is necessary that the second excitation step occurs before loss processes (collisional, intramolecular radiationless, and radiative) depopulate the intermediate A*, while with the non-resonant process the two photons must reach the absorber within about 10^{-15} s of each other. The resonant process has a higher probability than the non-resonant one, although the two formally distinct mechanisms merge if the photon energy of the exciting radiation in the non-resonant process approaches that of a real intermediate state. In both the sequential and the simultaneous mechanisms, the essential feature is that the energy of individual photons — still quantized according to the Planck relation — is stored in such a way that the energy of several photons can be used cooperatively.

2.38 The optical absorbance of a system absorbing multiple photons is dependent on the incident light intensity, i.e. the Beer–Lambert law (see Section 1.3) no longer holds. This behaviour is most obvious in the non-resonant, simultaneous process involving the virtual intermediate level. The system may be completely transparent at low light intensities, yet absorb radiation at the same wavelength when exposed to high intensities.

Multiphoton excitation in the ultraviolet region offers the photochemist several interesting opportunities. Excited states not normally accessible can be populated. For example, biphotonic excitation can be used to populate states with the same parity as the ground state (the selection rules are $g \leftrightarrow g$ and $u \leftrightarrow u$, contrary to the rules for single-photon absorption (see Section 1.6 and

Note 1.16). Of particular importance is the possibility of conducting high-energy photochemistry that is normally associated with ultraviolet radiation of wavelength shorter than, say, 190 nm (the 'vacuum' ultraviolet [Note 1.24]). The problems of experimentation in this wavelength region can be avoided by using multiphoton excitation with wavelengths in the 'conventional' ultraviolet region. Note 2.39 gives an example.

2.39 A typical example of multiphoton photochemistry in the vacuum ultraviolet region is the two-photon dissociation of CH_3I at $\lambda = 193$ nm

$$CH_3I + 2h\nu \rightarrow CH^* + H_2 + I$$

in which electronically excited CH^* radicals are formed. Ordinary ultraviolet dissociation of CH_3I yields $CH_3 + I$ as the fragments, and the experiments would be extremely difficult at the single-photon wavelength (~86 nm) corresponding to the biphotonic excitation.

Multiphoton ionization (MPI) is readily achieved with ultraviolet laser radiation. Compared with single-photon experiments, multiphoton excitation gives access to ionization studies on a vastly increased range of species. If resonant intermediate states are involved, the process is called *resonance-enhanced* multiphoton ionization (REMPI). The ions formed can be detected at very low concentrations, and MPI is of great value in multiphoton spectroscopic studies. MPI also provides a new perspective on photoionization–mass spectrometry, a technique in which wavelength-selected radiation allows highly specific ionization of particular species or of particular quantum states.

Infrared multiphoton absorption provides a unique source of molecules with high degrees of internal excitation. A new technique for the study of the problems of unimolecular dissociation is thus available. In addition, infrared multiphoton dissociation (IRMPD) generally yields products in their electronic ground states, which may well not be the case in conventional single-photon dissociation with ultraviolet or visible radiation (see Section 2.3 and Note 2.7). Infrared multiphoton dissociation gives products similar to those obtained by thermal dissociation or pyrolysis, but without the need for high temperatures of the bulk materials: it is thus a technique for conducting *vibrational photochemistry*. The technique has commanded much attention for the possibilities that it provides in isotope-selective chemistry and isotope separation (see Section 6.8).

2.13 State-to-state photochemistry

Much of this book is about electronically excited species populated above their thermal equilibrium populations, directly by the absorption of light or in subsequent processes. Considerable interest attaches to the nascent distribution of energy amongst electronic, vibrational, and rotational internal modes, and in translational motion, immediately following a photochemical event such as photodissociation [Note 2.40]. In a chemical reaction, the molecular system passes smoothly from reactants to products through an intermediate species; the field of reaction dynamics is concerned with how the laws of physical dynamics determine the approach of reactants and the departure of the products. In one sense, reaction dynamics is at the heart of all chemical transformations, and a better understanding of the dynamics represents a better understanding of the transformation itself. *State-to-state kinetics* provides some of the experimental information for theoretical approaches to reaction dynamics by attempting to investigate the rates of processes involving reactants in specific internal quantum states, and with specific velocities and

2.40 'Nascent' in the context of this discussion means that no significant redistribution of energy has taken place subsequent to the event, so that the disposition of energy amongst the various modes can be used to infer the dynamics of the event.

starting coordinates, to form products in equally well-defined quantum states, velocities, and angular distributions and momenta. Of course, most chemistry is not performed with state-selected species, but rather with something approaching thermally equilibrated statistical distributions. Nevertheless, these statistical distributions are made up of the individual states, so that the assembly of state-to-state processes makes up the whole reaction. Techniques are just now emerging that permit the 'high-resolution' study of state-to-state photochemistry, and we shall briefly mention some of the developments that will fuel physical photochemistry in the next few decades.

2.41 The experimental study of the nuclear motions of photodissociation offers a real challenge, because molecular structural changes occur over internuclear distances of a few tenths of a nanometre, corresponding to times on a femtosecond time-scale.

Photodissociation dynamics is a particularly promising part of reaction dynamics to study, because the interaction between a photon and a molecule is a 'half-collision', in which the initial properties of the photon are as well defined as possible. Theoretical models of the dissociation process can be tested against the results of experiments that measure the nascent energy and momentum disposal. The models frequently indicate that there is strong sensitivity to the quantum states in the absorbing molecule, so that the fullest experimental tests will require state-selected reactants as well as state identification in the products. The observable quantities that are most sensitive to the dynamics of dissociation seem to be the rotational energy disposal and the angular distributions and orientations of the fragments, and many of the more sophisticated studies have attempted to look at these parameters.

2.42 Investigation of the nascent energy disposal [Note 2.40] requires at least that intermolecular collisions do not redistribute the energy amongst the modes. Very low gas pressures may be needed, which is why this kind of work is essentially confined to gas-phase systems. Collision-free molecular beams may offer an even better way of avoiding collisions.

While intermolecular energy exchange can be minimized by avoiding collisions [Note 2.42], intramolecular energy exchange within the same molecule cannot. In these circumstances, the only way to obtain the nascent energy distributions is to probe the dissociating system at times short enough after the photon has been absorbed for energy exchange to be impossible. To meet this aim, and to understand the exchange processes themselves, very fast time-resolved experiments are required.

3 Photophysics

3.1 Emission and loss processes

Several of the pathways for loss of electronic excitation shown in Fig. 1.5 do not lead to any chemical change. Instead, the energy becomes used in a physical process such as the emission of light as *luminescence* (path (vi)), or becomes lost by *quenching* (*physical deactivation*, path (vii)). Alternatively, the excitation energy can populate a new level in the same or a different molecule by *intramolecular* or *intermolecular energy transfer* (path (v) or path (iv)); the new state or the new molecule excited by the energy transfer process can, of course, emit its own radiation. Luminescence, quenching, and the two types of energy transfer are important aspects of *photophysics*, and form the subject of this chapter.

Luminescent emission provides some of the most reliable information about the nature of primary photochemical processes. Competition exists between emission and other fates of excited species (quenching, reaction, decomposition, etc.), and the dependence of emission intensity on temperature, reactant concentrations, and other experimental parameters, may yield valuable data about the nature and efficiencies of the various processes. In particular, quenching by bimolecular collisions, and unimolecular energy degradation by radiationless transitions, are almost always best studied in terms of their effect on the intensity of luminescence. As well as possessing this fundamental interest, luminescent phenomena are also of considerable importance in several commercial and scientific applications (see, for example, Section 6.9).

The individual luminescent phenomena are named according to the mode of excitation of the energy-rich species. Emission from species excited by absorption of radiation is referred to as *fluorescence* or *phosphorescence*. Emission following excitation by chemical reaction (of neutral or charged species) is *chemiluminescence*, and is described briefly in Section 3.8.

3.1 There are other means of producing electronic excitation, for instance by using heat or an electric field, that do not really fall within the scope of this book.

The two emission processes in which the ultimate source of excitation is absorption of radiation — fluorescence and phosphorescence — were originally distinguished in terms of whether or not there was an observable 'afterglow'. That is, if emission of radiation continued after the exciting radiation was shut off, the emitting species was said to be phosphorescent, while if emission appeared to cease immediately, then the phenomenon was one of fluorescence. The essential problem is what is meant by 'immediately' in this context, since the observation of an afterglow will obviously depend not only on the actual rate of decay of the emission, but also on the techniques used to observe it.

The decay of intensity of luminescent emission is characterized quantitatively by the radiative *lifetime* (τ), which is the time taken for the intensity to decay to 1/e of its initial intensity immediately after the exciting radiation is shut off [Note 3.2 and Fig. 4.5]. Radiative lifetimes are usually measured in present-day experiments by exposing the luminescent sample to

3.2 τ is very simply related to the Einstein *A*-factor for emission (see Section 1.3 and Note 1.9), since $\tau = 1/A$. It is also very closely related to the radiative *half-life* (the time taken for the intensity to decay one-half of its initial value), since the half-life is roughly equal to $0.69/A$. These matters are discussed in more detail in Section 4.4.

a brief flash of light (see Section 1.10) and then following the decay of intensity with fast electronics that can record the intensity on a time scale of nanoseconds or even less. However, in the 1930s, with less sophisticated instrumentation, a luminescence with a lifetime of less than about 10^{-4} s was thought to be short lived and, hence, fluorescent. In 1935, Jablonski suggested that phosphorescence was emission from some long-lived metastable electronic state lying lower in energy than the state populated by absorption of radiation (see Section 1.9). This long-lived metastable state is now known to be the first excited triplet in many cases where the ground state of a molecule is a singlet. The long lifetime of the emission is a direct consequence of the 'forbidden' nature of a transition from a triplet to a singlet: that electric dipole transitions occur at all where $\Delta \mathbf{S} \neq 0$ is due to the inadequacy of $\mathbf{S}$ to describe a system in which there is spin–orbit coupling (see Note 1.17). Extension of this idea to other systems, not necessarily triplet–singlet, in which $\Delta \mathbf{S} \neq 0$ leads to the useful definition of phosphorescence as a *radiative transition between states of different multiplicities*: fluorescence is then understood to be a radiative transition between states of the *same* multiplicity [Note 3.3].

3.3 The definitions of fluorescence and phosphorescence given in the text are used almost universally by organic photochemists, although they might be extended to include within the scope of phosphorescence those emission processes involving a transition forbidden by *any* selection rule rather than just the $\Delta \mathbf{S} \neq 0$ rule. Since the distinctions between allowed and forbidden transitions are not sharp, the definitions lack some precision. Typically, the *A* factor for emission of a fully allowed optical transition is 10^8 to $10^9\,s^{-1}$, corresponding, very roughly, to radiative lifetimes (see Notes 1.9, 3.2 and Section 4.4) of 10^{-8} to 10^{-9} s. Forbidden transitions, on the other hand, can have lifetimes associated with them of seconds or more.

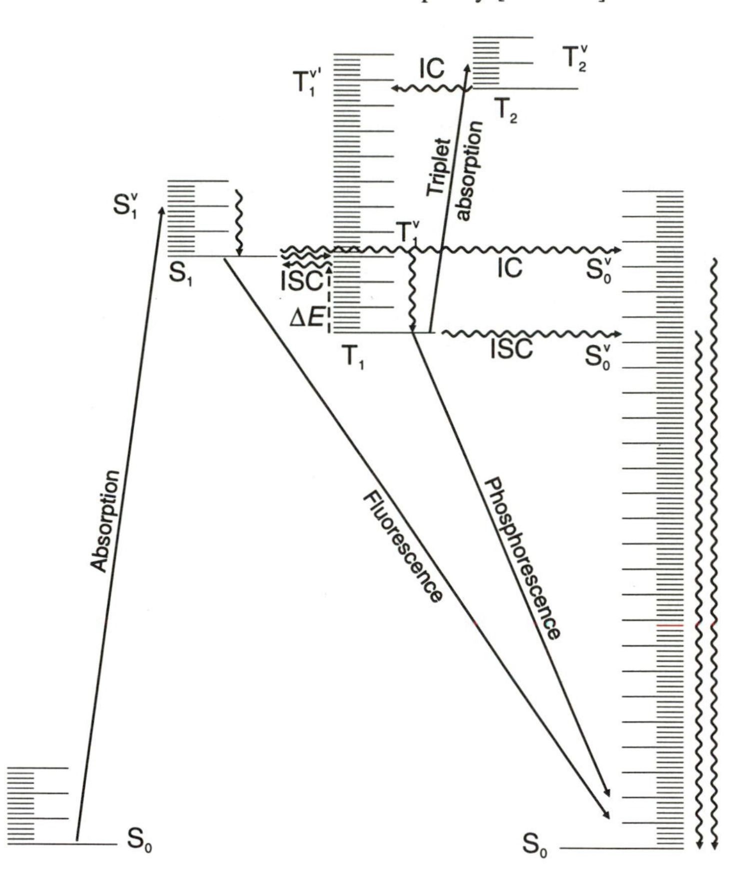

Fig. 3.1 Jablonski diagram showing absorption, and the emission processes of fluorescence and phosphorescence.

3.4 The Jablonski diagram does not attempt to represent the molecular shapes and sizes, and the vibrational levels drawn for each state do not usually correspond to the actual υ'', υ' numberings and spacings. On the other hand, the energies of the vibrational ground states of each electronic level are shown correctly if the experimental evidence is available.

In Section 1.9, we mentioned that a Jablonski diagram could be used to indicate the electronic and vibrational energies of complex molecules. Figure 3.1 is such a diagram. The S_0, S_1,..., T_1,... notation is employed. Wavy lines on the diagram represent radiationless energy conversion [Note 3.5]: the vertical wavy lines within a particular electronic state indicate degradation of vibrational excitation (probably by a collisional, *inter*molecular process), while the horizontal wavy lines indicate intramolecular energy exchange. Electronic energy exchange permitted by the $\Delta S = 0$ rule and that forbidden by it are labelled IC and ISC respectively (see Section 1.9).

3.5 Both internal conversion and intersystem crossing are assumed to take place with no change in total (electronic + vibrational) energy, and the wavy lines are therefore horizontal on the diagram (i.e. no translational or rotational energy is released in an intramolecular electronic energy exchange).

Absorption of radiation in a singlet–triplet transition is weak, since it is forbidden in the same way as the triplet–singlet phosphorescent emission. It follows that excitation of phosphorescence by direct absorption of radiation is inefficient, and phosphorescence is much more usually the result of emission from a triplet populated by intersystem crossing from an excited singlet formed on absorption. The sequence of events is illustrated in Fig. 3.1. Absorption populates S_1^v; vibrational energy, at least in condensed phases, is rapidly degraded and S_1^0 can then lose its energy by radiation, intersystem crossing (ISC) to T_1^v, or internal conversion (IC) to S_0^v. It is, perhaps, surprising that ISC to T_1^v, which is spin forbidden, can compete effectively with spin-allowed fluorescence and IC to S_0^v; phosphorescence is, however, observed in many systems, suggesting that IC from $S_1 \rightsquigarrow S_0$ is relatively inefficient. A complete understanding of the photochemistry of a molecule really requires that the efficiencies (i.e. quantum yields) be known for all the processes occurring. Even if chemical reaction, decomposition, and physical quenching of an excited species do not occur, it is still necessary to measure quantum yields for fluorescence (ϕ_f), phosphorescence (ϕ_p), intersystem crossing $T_1 \rightsquigarrow S_0$ (ϕ_{ISC}) and internal conversion $S_1 \rightsquigarrow S_0$ (ϕ_{IC}). If just these four processes occur, it follows that

$$\phi_f + \phi_p + \phi_{ISC} + \phi_{IC} = 1 \qquad (3.1)$$

Quenching or collisional deactivation

In addition to loss by emission and intramolecular energy transfer (IC and ISC), excited states often suffer losses in quenching processes. Quenching is especially important in the liquid phase where collisions are very frequent. At low pressures in the gas phase, collisional processes may not be competitive with emission or energy transfer, and in solids collisions may be hindered by the rigidity of the structure.

Quenching is a particular case of energy transfer between an excited molecule, A*, and a quenching species, usually represented as M

$$A^* + M \rightarrow A + M \qquad (3.2)$$

in which the excitation in the acceptor molecule (M) is not relevant. The electronic excitation in A* usually becomes 'degraded' to vibrational, translational, or sometimes rotational, energy in M. The important feature is that A* has lost its excitation (become 'deactivated'). Intermolecular energy transfer is discussed further in Section 3.5, but, for the present purposes, it is necessary only to note that the rate of the bimolecular process (3.2) is k_q[A*][M] where k_q is the rate constant for the process.

Quenching competes with emission, so that the presence or addition of quenching species reduces the intensity of fluorescence or phosphorescence. The kinetics of quenching are explored in Section 4.4, but there are some remarks that it is appropriate to make at this point. In a fluorescent process in which just the two processes of quenching (reaction (3.2)) and emission

$$A^* \rightarrow A + h\nu \quad (3.3)$$

occur, the quantum yield for fluorescence (ϕ_f) can be identified simply as the fraction of A* molecules that emit, and thus as the ratio of the number of molecules that emit to the total number that both emit or are deactivated. The rate of emission is, of course, A[A*]. Thus

$$\phi_f = A[A^*]/(A[A^*] + k_q[A^*][M]) = A/(A + k_q[M]) \quad (3.4)$$

Equation (3.4) can be used to generate one form of the *Stern–Volmer relationship*, because inversion of the equation yields the result

$$1/\phi_f = 1 + k_q[M]/A \quad (3.5)$$

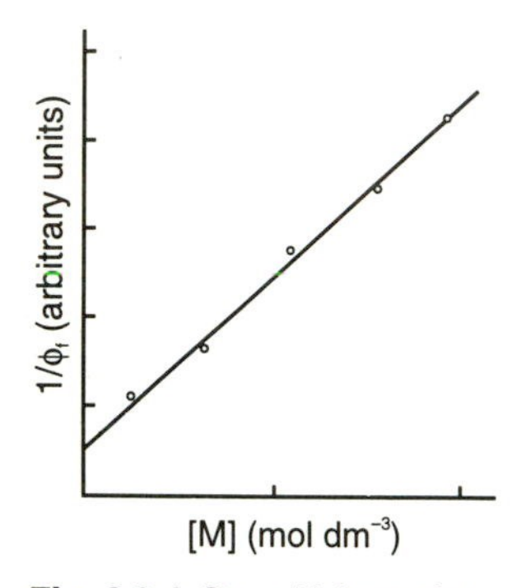

Fig. 3.2 A Stern-Volmer plot.

If $1/\phi_f$ is plotted against [M], as in Fig. 3.2, then a straight line should result of slope k_q/A, so that a knowledge of A allows determination of k_q. Usually, it is not ϕ_f itself that is determined, but rather the *intensity* of fluorescence that is measured, but the principle remains the same because the intensity is directly related to the quantum yield: the use of intensities in Stern–Volmer plots is discussed in Section 4.4.

In solution, the diffusive process limits the rate at which the excited species and the quencher come together, but prolongs each *encounter* so that several hundred collisions are possible before the two species diffuse apart (see also Section 2.5). Thus, even if not every molecular *collision* leads to deactivation, the rate of quenching may be determined more by the rate at which the quenching and excited molecules reach each other through diffusion than by the rate of collision between them. The rate of quenching is then said to approach the *diffusion-controlled limit* [Note 3.6].

3.6 An *approximate* expression for the diffusion-limited rate constant, k_{diff}, is given by the Debye equation

$$k_{diff} \sim \frac{8RT}{3\eta} \times 10^3\ \mathrm{dm^3\ mol^{-1}\ s^{-1}}$$

where η is the viscosity of the solvent in $\mathrm{N\,s\,m^{-2}}$. For water at room temperature, $\eta \sim 10^{-3}\,\mathrm{N\,s\,m^{-2}}$, so that $k_{diff} \sim 10^{10}\,\mathrm{dm^3\,mol^{-1}\,s^{-1}}$.

Intramolecular deactivation

It must be recognized that the radiationless intramolecular IC and ISC processes compete with emission of fluorescent or phosphorescent emission, just as the collisional processes do. Indeed, it is impossible to prevent the occurrence of these processes, while it may be possible to remove the influence of bimolecular quenching. Equation (3.1) embodies the idea that the *total* quantum yield for all the processes will be unity in the absence of quenching. The individual quantum yields, ϕ_f, ϕ_p, ϕ_{IC}, and ϕ_{ISC} are the ratios of the numbers of molecules decaying by the particular channel to the total number of molecules that lose their electronic excitation (equal, of course, to the total number of molecules excited). The IC and ISC processes, $S_1 \rightsquigarrow S_0$ and $S_1 \rightsquigarrow T_1$, thus reduce the intensity (and quantum yield) of fluorescence, while ISC $T_1 \rightsquigarrow S_0$ reduces the intensity of phosphorescence directly and IC $S_1 \rightsquigarrow S_0$ reduces it indirectly (by decreasing the population of S_1 from which the emitting T_1 level is excited). Efficient fluorescence or phosphorescence thus depend on the depopulating IC and ISC channels not competing with the emission process. Efficient phosphorescence also requires ISC $S_1 \rightsquigarrow T_1$ to be fast, and the other ISC and IC channels to be slow. The relative probabilities (that is, the rates) of these various intramolecular energy transfer processes therefore determine whether or not a molecule is likely to emit, and if the emission is more likely to be fluorescence or phosphorescence. The factors that control the rates of IC and ISC are examined in Section 3.4.

3.2 Fluorescence

An electronically excited atom must lose its energy either by emission of radiation or by collisional deactivation: chemical decomposition is not possible, and neither is radiationless degradation, which would require an increase in translational energy. At low enough pressures, therefore, fluorescent emission is expected from all atoms. Many molecular species, however, either do not exhibit fluorescence or fluoresce only weakly even when bimolecular reaction or physical deactivation do not occur. Some general principles can suggest whether a polyatomic organic molecule is likely to be strongly fluorescent. First, absorption must occur at a wavelength long enough to ensure that chemical dissociation does not take place. Absorption to an unstable state is clearly very unlikely to result in fluorescence. Further, in many molecules in which the absorption corresponds to an energy greater than the cleavage energy of the least stable bond, no fluorescence is observed. Secondly, intramolecular energy transfer must be relatively slow compared to the rate of radiation. This appears to mean that ISC from $S_1 \rightsquigarrow T_1$ must be slow (we have mentioned on p. 41 and shall discuss later in this section the inefficiency of the IC $S_1 \rightsquigarrow S_0$ process). Geometrical factors such as rigidity and planarity can also affect the efficiency of fluorescence (see Section 3.4).

3.7 In Section 3.4 we shall see that ISC is normally slower for $^1(\pi,\pi^*) \rightsquigarrow {}^3(\pi,\pi^*)$ states than for ISC involving (n,π^*) states, and that the efficiency of the process increases as the energy separation of S_1 and T_1 decreases. Simpler carbonyl compounds, in which the longest absorption corresponds to an $n \rightarrow \pi^*$ transition, are rarely fluorescent (but often phosphorescent) while aromatic hydrocarbons ($\pi \rightarrow \pi^*$ absorption) are frequently fluorescent. Increasing conjugation in hydrocarbons shifts the first $(\pi \rightarrow \pi^*)$ absorption maximum towards longer wavelengths, and thus increases the probability of fluorescence rather than decomposition. High ring density of the π electrons also seems important for high fluorescence yields.

There are two different types of spectrum discussed in connection with luminescence phenomena: the *excitation spectrum* and the *emission spectrum*. The first is obtained by measuring the intensity of the emission (which may or may not be spectrally resolved) as the excitation wavelength is altered, for example by scanning a monochromator. The emission spectrum is obtained by measuring the emission intensity as a function of wavelength for excitation at a fixed wavelength.

The detailed nature of the fluorescence emission spectrum is determined in part by whether the fluorescence is excited in a condensed-phase sample or in a gas, and, if in a gas, the pressure. The general ideas can be explained with reference to simple energy diagrams (Figs 3.3 (a)–(d) on the next page) showing schematically two electronic energy states and some vibrational levels associated with them.

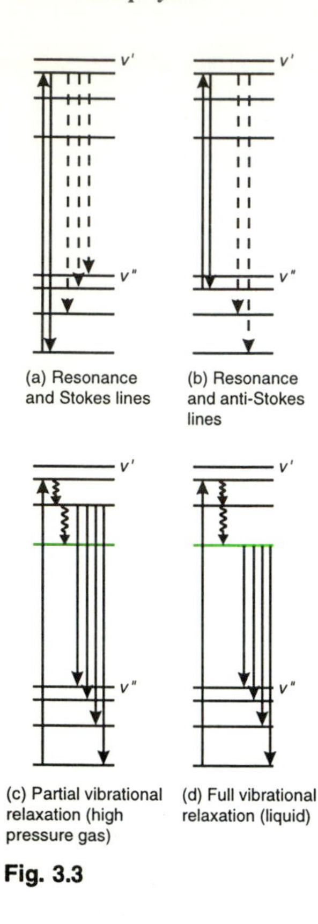

(a) Resonance and Stokes lines

(b) Resonance and anti-Stokes lines

(c) Partial vibrational relaxation (high pressure gas)

(d) Full vibrational relaxation (liquid)

Fig. 3.3

The simplest type of fluorescence is *resonance fluorescence*, in which the radiation emitted is of the same wavelength as the exciting radiation. Resonance fluorescence is observed only in the gas phase at low pressures, and only with atoms or simple molecules. Absorption of monochromatic light, of wavelength corresponding to a specific vibrational transition $(\upsilon', 0)$ populates exclusively the υ' level of the upper state and radiation from that state gives rise to the resonance fluorescence (see Fig. 3.3 (a)). Transitions also occur from υ' to υ'' levels higher than zero, so that a progression of bands is observed at wavelengths longer than the exciting wavelength, in accordance with an empirical observation of Stokes: the lines are called *Stokes lines* [Note 3.8]. Irradiation by polychromatic light can obviously excite many υ' levels, and fluorescent emission can then be observed from all these different levels.

Stepwise collisional relaxation of vibrational excitation is a relatively efficient process, requiring in the region of 1–100 collisions for many quenching gases. Resonance fluorescence is not expected, therefore, at pressures at which the kinetic collision frequency greatly exceeds the spontaneous emission rate, and, for $A \sim 10^8\,\mathrm{s}^{-1}$, observation of resonance emission is confined to pressures at least below about 1 mmHg (and less, if A is smaller than $10^8\,\mathrm{s}^{-1}$). Lower vibrational levels of the upper electronic state are populated from the υ' level produced on absorption, and at moderate pressures, at which emission and vibrational quenching still compete, emission may be observed from all vibrational levels in the upper state up to υ' (see Fig. 3.3 (c)).

At higher gas pressures, at which the collision rate greatly exceeds the rate of emission, vibrational relaxation is essentially complete, and no fluorescence is observed from $\upsilon' > 0$ (see Fig. 3.3 (d)). Vibrational relaxation is extremely probable in solution, and fluorescence from vibrationally excited levels is never observed in the liquid phase. Furthermore, neither the fluorescence spectrum nor deactivation rates are affected by changes in exciting wavelength so long as it lies within the absorption bands.

The intensity of each vibrational emission band depends, in the same way as absorption intensities, on the operation of the Franck–Condon principle. The $(0, 0)$ band is the most intense for many organic molecules, and the maxima of intensity both in absorption and in emission therefore correspond to the same transition. This observation suggests that upper and lower electronic states of such molecules must be of similar size and shape, and it is likely that the vibrational *spacings* will be the same in both states. In contrast, simple diatomic species frequently have greatly different internuclear separations in ground and excited states. Figure 3.4 (p. 45) indicates transitions in absorption and emission for two electronic states with similar vibrational spacings. Figure 3.5 (p. 45) shows emission and absorption spectra of a solution of anthracene in benzene. The two spectra are almost mirror images of each other on the wavenumber (i.e. energy) scale employed. Because the vibrational spacings are similar (see Fig. 3.4), the $(0, 1)$ emission band will lie at the same energy below the $(0, 0)$ band as the $(1, 0)$ absorption band lies above it, and

3.8 *Anti-Stokes lines*, much weaker than Stokes lines, are observed at shorter wavelengths than the exciting wavelength, and result from absorption from $\upsilon'' > 0$ and fluorescence to lower υ'' levels (see Fig. 3.3 (b)): thermal populations of $\upsilon'' > 0$ are small and the anti-Stokes lines are correspondingly weak.

so on. This mirror-image relationship is of frequent occurrence in the fluorescence of organic substances, and assumption of its existence is sometimes useful in sorting out overlapping emission spectra.

Organic fluorescence usually originates from the *lowest excited singlet* level, S_1, even though absorption may initially populate a higher singlet (e.g. S_2, S_3, . . . S_n). Rapid internal conversion occurs from S_n to S_1, followed by vibrational degradation. This behaviour is encapsulated in *Kasha's rule*, which states that *the emitting electronic level of a given multiplicity is the lowest excited level of that multiplicity*. One organic compound whose fluorescence is an exception to this rule is azulene, and many thioketones (RR′C=S) behave in a similar manner (see Note 3.16 for an explanation).

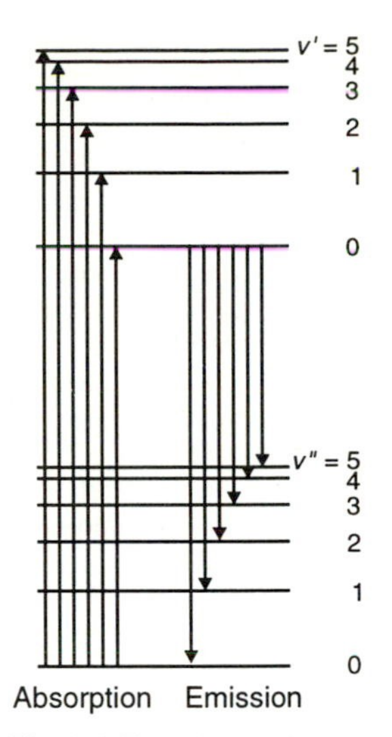

Fig. 3.4 Two electronic states with similar vibrational spacings.

Because fluorescent emission in organic compounds comes predominantly from the lowest vibrational level of the lowest excited singlet state of the molecule, it is often found that the fluorescence quantum yield is independent of the wavelength of the exciting radiation. Since, by definition, $I_f = I_{abs}\phi_f$, and the Beer–Lambert law can be use to calculate I_{abs},

$$I_f = I_{abs}\,\phi_f = I_0\,\phi_f\,(1 - e^{-\alpha Cd}) \tag{3.6}$$

At very low concentrations, where $\alpha Cd \ll 1$,

$$I_f = I_0\,\phi_f\,\alpha\,Cd \tag{3.7}$$

It follows from the constancy of ϕ_f with exciting wavelength that I_f is proportional to α at any wavelength if the incident intensity is the same at all wavelengths. That is, the *fluorescence excitation spectrum* is the same as the absorption spectrum in sufficiently dilute solutions. This result forms the basis of *spectrofluorimetry*, a technique that makes it possible to obtain 'absorption' spectra at very low solute concentrations (typically, at concentrations as low as 10^{-9} mol dm^{-3}); spectrofluorimetry can thus be a useful analytical tool.

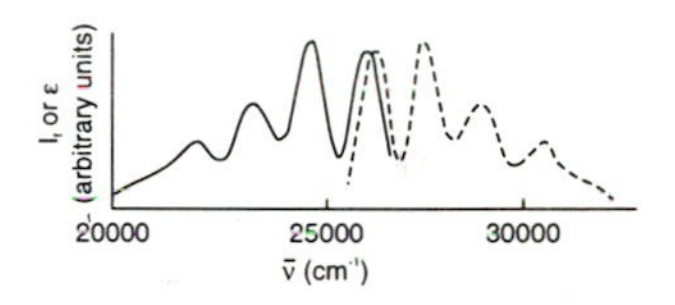

Fig. 3.5 Emission (solid line) and absorption (broken line) spectra of a solution of anthracene in benzene.

3.3 Phosphorescence

As explained in Section 3.1, phosphorescence is luminescence of a forbidden emission, and for most organic molecules with singlet ground states, this emitting species is a triplet. Because of the long radiative lifetime of phosphorescent transitions, collisional deactivation of the triplet competes effectively with radiation, and visible phosphorescence is not normally observed unless the collisional deactivation rate is sufficiently reduced. In rigid media, species are unable to diffuse towards each other, and bimolecular deactivation is slow. The earliest investigations of phosphorescence employed solutions of dyes in gelatin, and subsequently in boric acid glass at room temperature. More satisfactory rigid media are now used: mixtures of ether, isopentane, and ethanol (EPA) frozen at liquid nitrogen temperature (77 K) are frequently employed, and thin films of various plastics are becoming popular as rigid matrices. The highest purity of the solvents is necessary to avoid swamping the phosphorescence of the solute by luminescence of the impurities.

3.9 The spacing of emission bands indicates the vibrational levels in the ground electronic state, while the spacing of absorption bands depends on vibrational spacing in the upper state. See Fig. 3.4 for an explanation.

3.10 The (0,0) bands in Fig. 3.5 lie at slightly different wavelengths in absorption and emission. The separation results from energy loss to the solvent environment. Equilibrium interactions with the solvent may be different for ground and excited states of the solute.

3.11 There is direct evidence that the emitting state in phosphorescence is a triplet. Flash photolysis (see Section 1.10) can be used to produce high transient concentrations of triplet species: kinetic spectroscopy of absorption to a higher triplet permits the build-up and decay of the triplet level to be followed. The decay rates for the triplets and the phosphorescence are identical, confirming the connection between the two. Similar results are obtained from observations of triplets using electron paramagnetic resonance (EPR) spectra.

3.12 Spin–orbit perturbation is forbidden between states of the same configuration, so that, for example, a $^3(\pi,\pi^*)$ state must 'borrow' its singlet character from $^1(n,\pi^*)$ and $^1(\sigma,\pi^*)$ states rather than from $^1(\pi,\pi^*)$. Similarly, a $^3(n,\pi^*)$ state mixes with a $^1(\pi,\pi^*)$ state. Since radiative transitions from $^1(\pi,\pi^*)$ states to the ground state are fully allowed, while transitions from $^1(n,\pi^*)$ are, in general, somewhat forbidden, it follows that $^3(n,\pi^*)\rightarrow S_0$ transitions are more allowed than $^3(\pi,\pi^*)\rightarrow S_0$. Experimental observations confirm the predictions: in aromatic hydrocarbons having a $^3(\pi,\pi^*)$ state for T_1 the radiative lifetime is roughly 1–10 s, while for carbonyl compounds possessing a lowest triplet state of $^3(n,\pi^*)$ character the lifetime is usually 10^{-2}–10^{-1} s.

Although the first observations of phosphorescence were confined to rigid glasses, it was soon appreciated that phosphorescence could appear in other phases. Emission from biacetyl vapour is one of the best-known examples of gas-phase phosphorescence. Fluid solutions of species that are phosphorescent in low-temperature glasses also generally show emission, so long as the radiationless transitions from T_1 to S_0 do not show an increased rate at the higher temperatures. It is, of course, essential that the solvent does not deactivate the triplet, and that quenching impurities are rigorously excluded. Residual impurities may still make the emission intensity weak, and artificially reduce the luminescence lifetime. Perfluoroalkanes make suitable solvents for the study of phosphorescence at room temperature.

Absorption leading to direct population of an excited triplet state from the singlet ground state is weak because the transition is forbidden. However, in some cases it has proved possible to excite phosphorescence by irradiation with light absorbed in the $T_1 \leftarrow S_0$ system. Just as with fluorescence, there is often a mirror-image relationship between absorption and phosphorescence spectra. It would appear, therefore, that in relatively large organic molecules the vibrational spacings are nearly identical in all three lowest states (S_0, T_1, and S_1). The (0, 0) separations in the $T_1 \leftarrow S_0$ absorption and emission spectra are, however, relatively large ($\sim 500\,\text{cm}^{-1}$) as a result of slight conformational differences between ground and excited states. Hence, triplet energies based solely on the position of the (0, 0) band in emission may not represent faithfully the energetics of an absorbing system. Good absorption spectra of $T_1 \leftarrow S_0$ transitions are difficult to obtain by ordinary techniques, but the weakness of the absorption makes it possible to use the *phosphorescence excitation spectrum* to determine the absorption spectrum (this is *spectrophosphorimetry*, which is the analogue of spectrofluorimetry discussed at the end of Section 3.2).

Phosphorescence most commonly follows population of T_1 via intersystem crossing from S_1, itself excited by absorption of light. The T_1 state is usually of lower energy than S_1, and the long-lived (phosphorescent) emission is almost always of longer wavelength than the short-lived (fluorescent) emission. The relative importance of fluorescence and phosphorescence depends on the rates of radiation and intersystem crossing from S_1; the absolute quantum yields depend also on intermolecular and intramolecular energy-loss processes, and not only collisional quenching of T_1 but also intersystem crossing to S_0 are in competition with phosphorescent emission.

We must now give some consideration to the nature of triplet–singlet processes, both radiative and non-radiative, which are formally spin forbidden. Transitions can, however, occur with $\Delta\mathbf{S}\neq 0$ if $\mathbf{S}$ does not offer a good description of the system (see Note 1.17). Thus, transitions between triplet and singlet states can take place if the triplet has some singlet character, or vice versa. In organic molecules some 'mixing' of singlet and triplet states takes place as a result of a small amount of spin–orbit interaction. The relative probability of triplet–singlet transitions from (n,π^*) and (π,π^*) states is explained in Note 3.12. These remarks apply equally to both the phosphorescent emission process, $T_1 \rightarrow S_0 + h\nu$, and the ISC process that populates T_1 in the first place ($S_1 \rightsquigarrow T_1$). This ISC energy transfer is clearly of fundamental importance in phosphorescence, and having examined the basic ideas of phosphorescence, we should now turn our attention again to radiationless transitions in general.

3.4 Intramolecular energy transfer (IC, ISC) revisited

There are two distinct factors that determine the efficiency of the radiationless energy transfer processes (IC and ISC). The first depends on the electronic states, and describes the intrinsic probability of transition from one state to the other. The second factor is concerned with the vibrational levels, and makes itself felt through efficiencies that greatly increase as the electronic energy separation between $\upsilon = 0$ levels decreases.

Electronic factors

On the basis of the ideas just presented in Note 3.12, El-Sayed has suggested the following 'rules' for spin-forbidden intramolecular energy transfer

$$^{1\text{ or }3}(n,\pi^*) \longleftrightarrow {}^{3\text{ or }1}(\pi,\pi^*);\ {}^{3}(n,\pi^*) \not\longleftrightarrow {}^{1}(n,\pi^*);\ {}^{3}(\pi,\pi^*) \not\longleftrightarrow {}^{1}(\pi,\pi^*) \quad (3.8)$$

The rate of these transitions may be perturbed by the external environment. Such an influence is seen in the effects of the addition of paramagnetic molecules to the solvent [Note 3.13]. Heavy atoms in an environment also increase the probability of $S \leftrightarrow T$ radiative and radiationless transitions by inducing appreciable spin–orbit coupling in the solute [Note 3.14].

*Intra*molecular perturbation of transition probabilities is also important. For example, substitution in naphthalene by one iodine atom increases the transition probability for optical emission by a factor of nearly 10^4 and the rate of $T_1 \rightsquigarrow S_0$ ISC by nearly 1000. Furthermore, ϕ_p/ϕ_f increases by more than 8000, mainly because of the increased probability of $S_1 \rightsquigarrow T_1$ ISC in the substituted molecules.

Strong intramolecular perturbations may also arise when certain metal ions are chelated to an organic molecule. It is interesting that the natural porphyrins chlorophyll and haemin display markedly different photochemical behaviour, despite the similarities in their structures: chlorophyll has diamagnetic magnesium as its central ion, while haemin has paramagnetic iron (see Fig. 5.7 for the structure of chlorophyll).

3.13 Although O_2 and NO decrease phosphorescence yields because of their participation in efficient bimolecular quenching, they increase the rates both of the forbidden optical transition and of ISC. A dramatic demonstration of the increase in $T \leftarrow S$ absorption is afforded by pyrene solutions, which are normally colourless but which turn deep red in the presence of high pressures of oxygen.

3.14 Solutions of anthracene and some of its derivatives become less fluorescent on addition of bromobenzene, while the triplet–triplet absorption intensity increases as a result of enhanced $S_1 \rightsquigarrow T_1$ ISC.

3.15 Substitution affects the photochemistry of a species not so much through changes in energy levels (for example, the first triplet and excited singlet levels lie at 21,300 cm^{-1}, 32,000 cm^{-1}, respectively, in naphthalene, and shift only to 20,500 cm^{-1}, 31,000 cm^{-1} in 1-iodonaphthalene), as via changes in the relative probabilities of fluorescence, phosphorescence, and the IC and ISC processes.

Vibrational Factors

The probability for energy transfer can be shown experimentally to be some inverse function of the energy gap between the two states for a given type of electronic transition. This result can be understood in terms of the operation of the Franck–Condon principle in radiationless transitions. As explained in Section 1.9, in the radiationless case the principle demands that the transition

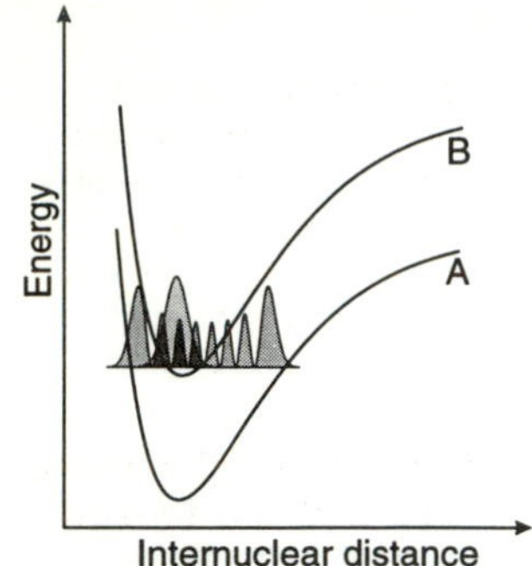

(a) Similar geometry, with large energy separation.

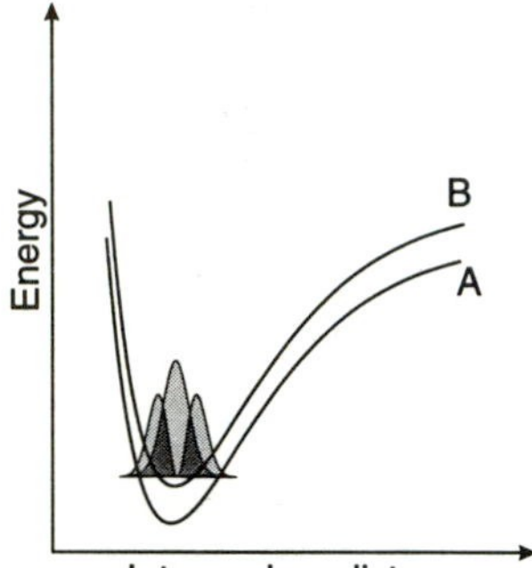

(b) Similar geometry, with small energy separation.

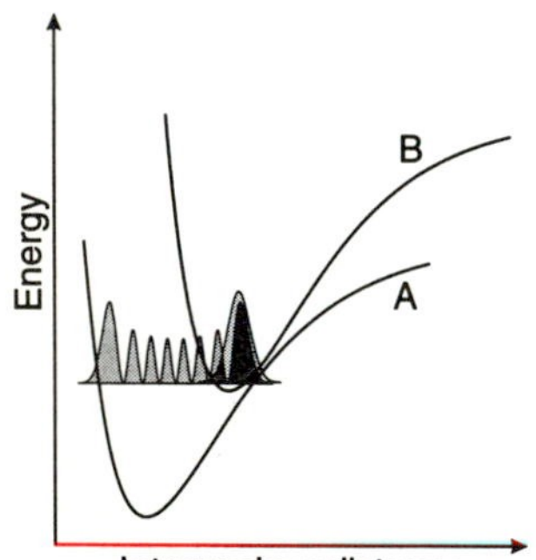

(c) Different geometry, with large energy separation.

Fig. 3.6 The Franck–Condon principle in radiationless transitions.

is 'horizontal' as well as 'vertical', so that it is confined to a very small region of a potential energy curve or surface. For a given probability of electronic transition, the overlap in the region between the vibrational probability functions for starting and finishing states will then determine the efficiency of energy transfer. Figure 3.6 extends Fig. 1.4 by illustrating three possibilities: the curves given may be regarded as potential energy curves for diatomic molecules, or as cross-sections through surfaces for more complex species. In Fig. 3.6(a), the two states, A and B, are of similar geometry, but of widely different energy. The lowest vibrational level, $\upsilon'=0$, in B has the same total energy as high υ'' in A. Because of the nature of the vibrational probability distributions, the overlap is small. In Fig. 3.6(b), however, the energy gap is much smaller, and the difference in vibrational quantum numbers υ' and υ'' is also smaller, with the result that there is far greater vibrational overlap. Thus, the efficiency of crossing will increase as $\upsilon''\rightarrow 0$; i.e. crossing to a state will be favoured if the state can be populated near $\upsilon''=0$, which means that the *electronic* energy gap itself must be small. Only if the geometries of A and B are different, as in Fig. 3.6(c), can there be rapid radiationless conversion between two states of widely separated electronic energies. In general, energy separations lie in the order $(S_1{-}S_0) > (T_1{-}S_0) > (S_1{-}T_1)$, and the rate constants usually lie in the reverse order, even though the second two processes are formally spin forbidden. Molecular rigidity favours efficient fluorescence of a species (see p. 43). Without that rigidity, changes in geometry may occur, and $S_1 \rightsquigarrow S_0$ IC *can* depopulate S_1. The sum of ϕ_f and ϕ_{ISC} is often close to unity, so that the quantum yield for internal conversion, $\phi_{IC}=1-(\phi_f+\phi_{ISC})$, must be zero within experimental error. The observation made in Section 3.2 that organic fluorescence usually originates from S_1 rather than from higher singlets (S_n) is explained by the size of the $S_n{-}S_1$ and $S_1{-}S_0$ energy gaps. The former gap is much smaller, so that the $S_n \rightsquigarrow S_1$ IC process is much faster than $S_1 \rightsquigarrow S_0$, and S_n will not survive to emit radiation [see Note 3.16 on p. 49].

The same arguments apply to the rate of the $T_1 \rightsquigarrow S_0$ (ISC) process. As a general rule, the larger the molecule, the smaller $\Delta E(T_1{-}S_0)$, and ϕ_p is correspondingly low in large molecules.

The differences in the photochemistry of species whose absorption system is predominantly π–π^* or n–π^* are at least in part a result of the increased likelihood of $S_1 \rightsquigarrow T_1$ ISC for $S_1={}^1(n,\pi^*)$. Three factors work towards this increased likelihood of radiationless transition for ${}^1(n,\pi^*)$ states. First, both radiative and radiationless transitions to the S_0 ground state are partially forbidden, so that the ${}^1(n,\pi^*)$ state survives longer than a ${}^1(\pi,\pi^*)$ state and has more chance of undergoing ISC to T_1. Secondly, T_1 *may* be ${}^3(\pi,\pi^*)$, to which ISC from ${}^1(n,\pi^*)$ is favoured [Note 3.12]. Thirdly, the energy gap, $\Delta E(S_1{-}T_1)$ is often small compared to the gap from ${}^1(\pi,\pi^*)$ states [Note 3.17, p. 49]. In many carbonyl compounds, phosphorescence is strong, and fluorescence is weak or non-existent (typically $\phi_p/\phi_f > 1000$) because of efficient crossing from S_1 to T_1; a fairly large fraction of the molecules reaching T_1 phosphoresce, the magnitude of the $T_1\rightarrow S_0$ gap partly determining the rate of radiationless,

$T_1 \rightsquigarrow S_0$ ISC, decay. In hydrocarbons, S_1 is likely to be $^1(\pi,\pi^*)$, so that the transition to $^3(\pi,\pi^*)$ is strongly forbidden. Furthermore, $\Delta E(S_1–T_1)$ is large. The two factors together make $S_1 \rightsquigarrow T_1$ ISC less probable, and fluorescence becomes important in such hydrocarbons.

3.16 The peculiar behaviour of azulene (see p. 45), which shows fluorescence from S_2, can probably be explained by an extension of the ideas presented in the text. In the azulene molecule, the $S_2–S_1$ gap is relatively large, so that the normally rapid $S_2 \rightsquigarrow S_1$ conversion is slowed down, and fluorescence is mainly of the $S_2 \rightarrow S_0 + h\nu$ transition.

3.5 Intermolecular energy transfer

*Inter*molecular exchange of energy between two discrete partners is another important photochemical step. The 'acceptor' (A), which receives excitation from the 'donor' (D), then participates in those processes open to it as an electronically excited species. *Photosensitized* phenomena, such as sensitized reactions (see Section 2.6) and sensitized fluorescence or phosphorescence (see Section 3.6), in which the change of interest occurs in a species other than the one that absorbed radiation, are believed to be of great significance in photobiology; they also provide valuable insights into photophysical processes.

Exchange of energy between two different species is not so restricted with regard to exact equivalence of internal energy between initial and final states as in the case of intramolecular exchange, since any energy excess can be taken up by translation (or, more rarely, a deficiency can be supplied by the kinetic energy of collision). Energy exchange can take place between electronic, vibrational, rotational, and translational energy modes; except in the rare case of exact energy resonance between the internal modes, some energy is always converted to or from translation. We have already given implicit consideration to the degradation of electronic excitation to vibration, rotation or translation, in our discussion of physical quenching of luminescence. Much of the following discussion will, however, be directed towards electronic–electronic energy exchange, and it will be assumed that *excess* energy goes into other modes of excitation. Where the absorption spectrum of the acceptor overlaps the emission spectrum of the donor, there are, of course, quantized vibronic levels of A and D for which energy exchange is isoenergetic, and no change in translational energy is needed.

3.17 The splitting for singlet to triplet (π,π^*) is often from about 10,000 cm^{-1} to 20,000 cm^{-1}, while that for (n,π^*) states is usually only 1,500 cm^{-1} to 5,000 cm^{-1}.

Several different mechanisms of electronic energy transfer are believed to operate under different circumstances. They include: (i) collisional 'short-range' transfer; (ii) coulombic 'long-range' transfer; (iii) radiative transfer, and (iv) exciton migration. We shall discuss only the first two mechanisms, as they are the most important in solution-phase photochemistry.

Short-range, collisional, energy transfer

'Short-range' energy transfer arising from *exchange interaction* occurs over intermolecular or interatomic distances (henceforth referred to as r) not much exceeding the collision diameter; the interaction decreases in a complex fashion with r raised to a high power.

Energy transfer by exchange interaction may be thought of as a special kind of chemical reaction in which the chemical identity of the partners A and D does not change, but in which excitation is transferred from one to the

3.18 Multiplicities of transition states for given multiplicities of two separated species.

Separated species	Transition state
S + S	S
S + D	D
S + T	T
D + D	S + T
D + T	D + Q
T + T	S + T + Qn

where S = singlet, D = doublet, T = triplet, Q = quartet, and Qn = quintet.

3.19 Rate constants ($M^{-1} s^{-1}$) for triplet energy exchange.

Donor	Acceptor	ΔE (cm^{-1})	Rate Constant
Acet	Naph	−4200	1×10^{10}
Benz	Naph	−2800	1×10^{10}
Trip	Naph	−2100	2×10^{9}
Naph	Biac	−1700	9×10^{9}
Brom	Biac	−1000	3×10^{9}
Biac	Brom	1000	3×10^{7}
Biac	Naph	1700	2×10^{6}
Naph	Trip	2100	$<10^{4}$
Naph	Benz	2800	$<10^{4}$

Where
Acet = Acetophenone,
Benz = Benzophenone,
Trip = Triphenylene
Naph = Naphthalene
Brom = Bromonaphthalene
Biac = Biacetyl

other. The transition state is then expected to possess a separation between A and D not greatly different from the sum of the gas-kinetic collision radii, and energy transfer by the exchange mechanism is probably important only for values of r of this order. In common with other chemical processes, energy transfer can be efficient only if the process is adiabatic (see p. 19 and Note 2.32). Most chemical reactions involving ground-state partners can occur adiabatically, but in processes such as energy exchange, where several electronic states are involved, the requirement for adiabatic reaction may impose some restrictions on the possible states for A, A* and D, D* if there is to be efficient transfer of excitation. In atoms or small molecules there must be correlation of electron spin, orbital momentum, parity, and so on. However, correlation in complex molecules of low symmetry usually only involves the electron spin. To test for correlation, the possible total spin(s) of the transition state is calculated from the individual spins of the reactants by the addition of the quantum vectors (for example, $\mathbf{S}_1 = 1$ adds to $\mathbf{S}_2 = 1$ to give the resultant $\mathbf{S}_T = 2$, 1 or 0). It is then necessary to see if the product pair is also able to give one (or more) transition state(s) of the same spin as the transition state(s) reached from the reactants. Note 3.18 shows multiplicities in the transition state that can arise from some multiplicities of two separated species. Thus we can see, by use of the table, that all of the processes

$$\mathrm{A(T)+B(T)} \leftrightarrow \begin{array}{l} \mathrm{[S]} \leftrightarrow \mathrm{C(S) + D(S)} \\ \mathrm{[S\ or\ T]} \leftrightarrow \mathrm{C(D) + D(D)} \\ \mathrm{[S\ or\ T]} \leftrightarrow \mathrm{C(T) + D(T)} \end{array} \tag{3.9}$$

can occur adiabatically. On the other hand, the reaction

$$\mathrm{A(S) + B(T)} \not\leftrightarrow \text{[no common multiplicity]} \not\leftrightarrow \mathrm{C(S) + D(S)} \tag{3.10}$$

cannot proceed (in either direction) adiabatically. The rules derived from these arguments are known as the *Wigner spin correlation rules*.

The exchange of energy between molecules is most efficient if the amount of translational energy that must be liberated is small; thus we may expect rapid exchange of energy if vibronic levels are in near resonance. Exact matching of electronic (E) + vibrational (V) energy, such as is required in intramolecular energy transfer (see p. 48), is not demanded. Furthermore, with complex molecules possessing many vibrational modes, it is usually not difficult to find a near match in the total of E + V energies. Note 3.19 shows measured rate constants for several triplet–triplet energy exchange processes

$$\mathrm{D^*(T_1) + A(S_0) \rightarrow D(S_0) + A^*(T_1)} \tag{3.11}$$

and lists the energy difference between $D^*(T_1, \upsilon=0)$ and $A^*(T_1, \upsilon=0)$. A negative ΔE in the table represents an exothermic reaction. So long as the reaction is exothermic, energy transfer is fast (it approaches the diffusion-controlled limit: see Note 3.6), but when the reaction becomes endothermic, the rate rapidly drops off [Note 3.20, p. 51]. Similar results are obtained for singlet–singlet exchange.

Triplet–triplet energy transfer is sometimes treated as though it were a different phenomenon from singlet–singlet transfer. However, so far as the exchange interaction mechanism is concerned, the fact that both A and D change their spin multiplicity is of no account, since the reaction is adiabatic. Observed differences in photochemical behaviour arise from the much longer radiative lifetimes of triplet states.

3.20 The importance of the exothermicity of the transfer process is shown in the quenching of biacetyl phosphorescence by only those quenchers whose triplet level lies below that of biacetyl. The implication is that quenching involves triplet–triplet energy exchange, and subsequent experiments have, in fact, detected quencher triplets by their absorption spectra.

Long-range, coulombic, interactions

For many energy transfer processes, the interaction takes place when the partners are separated by more than the sum of the gas-kinetic collision radii [Note 3.21]. 'Long-range' transfer occurs by a direct mechanism involving electrical, or coulombic, interactions between transition dipoles [Note 3.22]. These dipoles are the ones involved in optical interactions with the electric vector of radiation: the usual optical selection rules apply to both the transitions $D^* \rightarrow D$ and $A \rightarrow A^*$. Higher multipoles can also interact, but dipole–dipole interactions are stronger than dipole–quadrupole interactions, and so on. The measured rate constants for transfer of excitation by the coulombic mechanism may greatly exceed the diffusion-limited rate constant, and not depend on solvent viscosity. This behaviour contrasts markedly with that observed for short-range transfer. (Note that k_{diff} is dependent on viscosity, as indicated by Note 3.6).

3.21 Energy transfer between excited singlet states of hydrocarbons has been shown to occur at distances between exchanging molecules of about 5 nm, or about 10 times the collision diameter.

Comparison of selection rules for intermolecular energy transfer

The efficiency of energy transfer occurring by the exchange interaction mechanism is related to whether the process can take place adiabatically, but *not* to whether optical selection rules permit radiative transitions in both donor and acceptor independently; this behaviour is one way in which exchange interaction and long-range coulombic interaction may be distinguished. For example, in the exchange interaction excitation of triplet states by triplet benzophenone, the efficiency of energy transfer is roughly the same both for naphthalene and for 1-iodonaphthalene. We saw on p. 47 that the $T_1 \rightarrow S_0$ radiative transition is more probable by a factor of nearly 10^4 in the substituted molecule, so that, in this case, the optical transition probability in the naphthalene molecule does not seem to affect the probability of energy transfer to it.

3.22 For a dipole–dipole interaction, theory predicts that the strength of the interaction should fall off as $1/r^6$, and relatively long-range energy exchange becomes possible.

According to the spin-selection rule, $\Delta S = 0$, long-range coulombic transfer should be impossible for any process involving multiplicity changes, and long-range triplet–triplet energy transfer would then be excluded. However, to the extent that spin–orbit coupling allows electric dipole optical transitions with $\Delta S \neq 0$ in complex molecules, coulombic transfer *can* occur by the dipole–dipole mechanism.

3.6 Sensitized and delayed fluorescence

In Section 2.6, we saw that chemical reaction can follow intermolecular energy exchange. Here we examine emission following such energy transfer between donor and acceptor species.

In 1922 Franck predicted that electronic excitation could be exchanged between atoms, and Cario and Franck subsequently demonstrated *sensitized fluorescence* in a mixture of mercury and thallium vapours. The mixture was irradiated with the $\lambda = 253.7$ nm resonance line of mercury [Note 3.23], to which thallium vapour is transparent, and emission was observed from the thallium. Absorption of light by Hg raises it to the resonance level, $^3P_1$, and energy is then transferred to the thallium

3.23 In Hg, the ground state is $^1S_0$. The first excited state is $^1P_1$, and the resonance transition is, strictly speaking, $^1P_1 \rightarrow {}^1S_0$ at $\lambda = 184.9$ nm. However, spin–orbit coupling is strong in Hg, so that the $^3P_1 \rightarrow {}^1S_0$ transition ($\lambda = 253.7$ nm) is intense; it is thus common practice to refer to this line as the resonance one.

$$\mathrm{Hg} + h\nu_{\lambda=253.7\,\mathrm{nm}} \rightarrow \mathrm{Hg^*} \qquad (3.13)$$
$$\mathrm{Hg^*} + \mathrm{Tl} \rightarrow \mathrm{Hg} + \mathrm{Tl^*} \qquad (3.14)$$
$$\mathrm{Tl^*} \rightarrow \mathrm{Tl} + h\nu \qquad (3.15)$$

Very many examples of sensitized fluorescence and sensitized phosphorescence are now known.

Energy may sometimes be transferred from an excited species to an acceptor that is already excited, thus raising the acceptor to a higher electronic state; the process is referred to as energy-pooling. Molecular energy-pooling has been established in many systems. Triplet–triplet pooling to give an excited singlet state is most common, partly because the relatively long lifetime of excited triplets favours the rare triplet–triplet bimolecular process. The reaction

$$\mathrm{D^*(T)} + \mathrm{A^*(T)} \rightarrow \mathrm{D(S)} + \mathrm{A^{**}(S)} \qquad (3.16)$$

is adiabatic with respect to spin, and for many organic molecules the first excited singlet is energetically accessible by pooling of energy from two triplets.

Both donor and acceptor can be molecules of the same chemical entity, so that reaction (3.16) provides a means of reaching the singlet state when only triplets are present in the system. Energy-pooling between two triplets is known as 'triplet–triplet quenching' or 'triplet–triplet annihilation', and is a mechanism for the emission of *delayed fluorescence*. For example, in anthracene the decay of fluorescence has two components, one with the normal fluorescence lifetime and the other slow, although the spectral distribution of both components is identical. The excitation mechanism (omitting radiationless decay or quenching steps) seems to be

$$\mathrm{A(S_0)} + h\nu \rightarrow \mathrm{A^*(S_1)} \qquad (3.17)$$
$$\mathrm{A^*(S_1)} \rightarrow \mathrm{A(S_0)} + h\nu \quad \text{normal fluorescence} \qquad (3.18)$$
$$\mathrm{A^*(S_1)} \rightsquigarrow \mathrm{A^*(T_1)} \quad \text{ISC} \qquad (3.19)$$
$$\mathrm{A^*(T_1)} \rightarrow \mathrm{A(S_0)} + h\nu \quad \text{normal phosphorescence} \qquad (3.20)$$
$$\mathrm{A^*(T_1)} + \mathrm{A^*(T_1)} \rightarrow \mathrm{A^*(S_1)} + \mathrm{A(S_0)} \quad \text{energy-pooling} \qquad (3.21)$$
$$\mathrm{A^*(S_1)} \rightarrow \mathrm{A(S_0)} + h\nu \quad \text{delayed fluorescence} \qquad (3.22)$$

Reactions (3.18) and (3.22) are, of course, identical, and are written twice to show the sequence of events leading to prompt and delayed fluorescence. Triplet-annihilation delayed fluorescence is sometimes known as *P-type delayed fluorescence* because it is observed in solutions of pyrene [Note 3.24].

3.24 Delayed fluorescence in pyrene shows an additional feature in that the delayed emission appears to derive mainly from the excimer $PP^*(S_0S_1)$ [P=pyrene], while the normal, prompt, fluorescence at moderate concentrations shows both monomer and excimer bands. Excimers will be discussed further in Section 3.9.

Another mechanism for generating long-lived fluorescent emission is *E-type delayed fluorescence*, named after the dye eosin. E-type delayed fluorescence shows spectral features characteristic of the normal, short-lived, fluorescence. However, the emission decays at the same rate as phosphorescence; further, no emission is observed at low temperatures, and there is an activation energy for the process. This kind of delayed fluorescence arises from thermal activation of S_1 from $T_1(\upsilon=0)$; the rate of activation is slow compared to the rate of loss of either T_1 or S_1, so that the decay of delayed fluorescence is determined by the decay of T_1. Figure 3.7 illustrates this excitation mechanism. The activation energy for the emission (calculated from the temperature dependence of intensity) should be identical to $\Delta E(S_1–T_1)$ (obtained spectroscopically) and, for many molecules, a remarkable agreement is indeed observed.

The P-type delayed fluorescence displayed by anthracene does not show the same dependence on temperature as the thermally activated E-type delayed fluorescence, and it may be distinguished from it by this means. A better diagnostic feature is the relation between the emitted and absorbed intensities, which is linear in E-type delayed fluorescence, but squared in the triplet-annihilation process. In addition, E-type delayed fluorescence has the same decay lifetime as that of the triplet–singlet phosphorescence in the same solution; delayed fluorescence excited by the triplet-annihilation mechanism should have a lifetime of about one-half of that of the phosphorescence, because of the second-order dependence on triplet concentration.

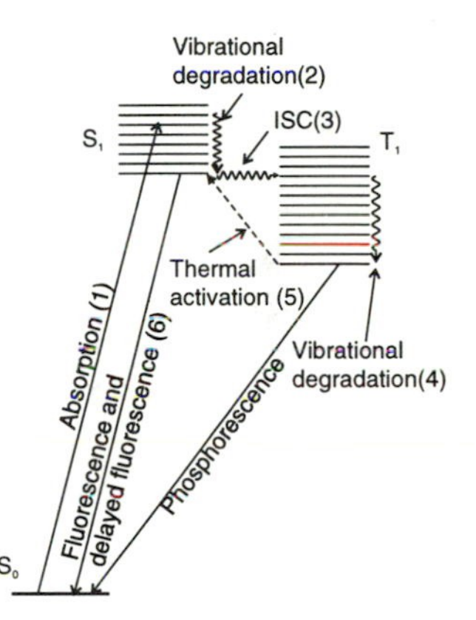

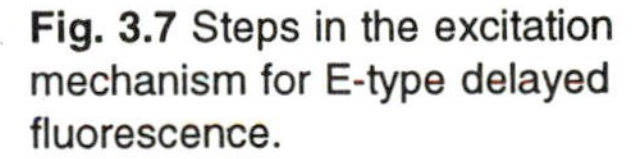

Fig. 3.7 Steps in the excitation mechanism for E-type delayed fluorescence.

3.7 Lasers

Emission from electronically excited species forms the basis for an important class of light sources, that of lasers. The properties of laser radiation have been referred to in Note 1.26. The term laser is an acronym for **L**ight **A**mplification by **S**timulated **E**mission of **R**adiation, and, as the name suggests, the operation of a laser depends upon the net production of stimulated emission (see Section 1.3). Absorption and stimulated emission both necessarily take place in systems containing electronic levels between which transitions occur, the rates of each process being determined by the population of the lower (absorption) or upper (emission) levels. *Net* emission (the difference between stimulated emission and absorption), and thus amplification of light, can only be achieved if there are more species in the upper state than in the lower state, a situation described as a *population inversion.*

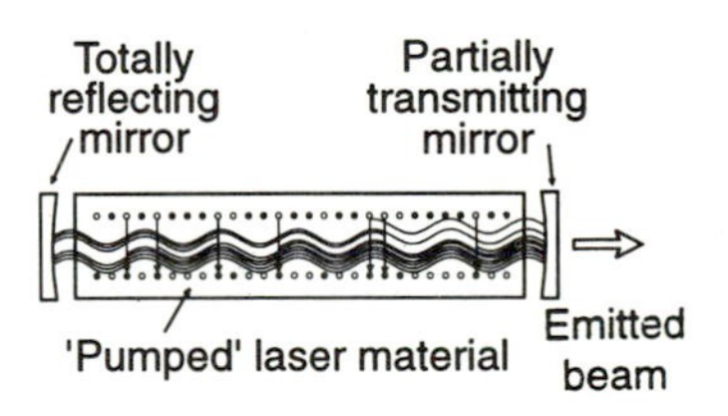

Fig. 3.8 A laser system. Open circles represent the upper state of the emitting species, and solid circles the lower state.

To understand the principles of laser action, it is helpful to consider a typical laser set-up. The essential features of one type of laser system are shown in Fig. 3.8. The species with the inverted population (the *lasing*

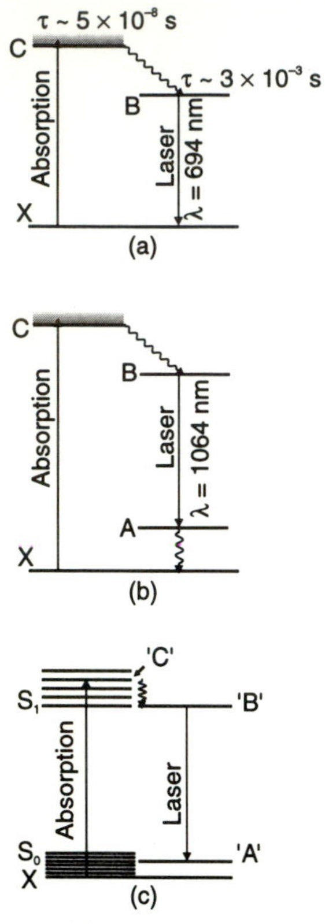

Fig. 3.9 Schemes for obtaining population inversion in optically pumped systems: (a) 3-level (ruby laser); (b) 4-level (Nd laser); (c) 4-level (dye laser).

medium) is contained within a cylindrical tube with a mirror at each end. One of these mirrors is totally reflecting, whilst the other is partially transmitting to enable some of the light produced to be harvested. The laser cavity is made to be resonant at the wavelength of the radiation, so that as the radiation traverses the medium the intensity increases through constructive interference. In the excited system, some of the excited species will lose their energy by spontaneous emission. The photons produced in this way can then interact with further excited species to stimulate the emission of more photons, and so on, thus greatly amplifying the radiation; spontaneous emission *seeds* the process of stimulated emission. The light amplification can continue until the population inversion is destroyed, so that the light is produced in *pulses* as population inversions are created and destroyed. Lasers in which the population inversion is maintained, by constantly replenishing the supply of excited species, are called *continuous-wave* (*CW*) lasers.

There are various approaches that may be adopted to achieve the population inversion necessary for laser action. Many of these rely upon the intra- and inter-molecular energy transfer processes that were introduced in Chapter 1 and are discussed further in this chapter. Thermal excitation cannot, by definition, lead to inversion in an equilibrated system. Nor can direct absorption of light give an inversion in a simple system consisting of two levels, because the incident ('pumping') radiation will not only excite the lower state to the upper, but also promote stimulated radiation from the upper state to depopulate it. Population inversion, and hence laser action, can, however, be achieved when three or more energy levels are involved. Figure 3.9 (a) shows a *three-level* system; one of the first lasers developed, the ruby laser, operates on such a mechanism. The third level, B, is populated by radiationless transition from the state initially populated by absorption (C). As long as this transition is rapid, and the radiative transition from B to X is relatively slow, a population inversion can be built up. Even so, more than 50% of the ground-state species must have been removed by pumping in order to achieve a population inversion. A potentially more efficient laser system uses four levels (Fig. 3.9 (b)); neodymium-based lasers, for example, use this system. As soon as state B is populated, there is an inversion with respect to level A, which is initially not populated at all. CW, as opposed to pulsed, operation requires, of course, that the level A is rapidly depopulated.

Organic dye lasers are effectively four level devices: although there may only be two electronic levels involved, the system makes use of the vibrational sub-levels and can therefore be thought of as using four levels (Fig. 3.9 (c)). As the dyes are usually in solution, collisional quenching is sufficient to maintain the population inversion. The lasers are tunable as a result of using systems which exhibit broad-band absorption and emission characteristics, from which the desired frequency may be selected (using, for example, a diffraction grating).

Gas-phase lasers often use electrical excitation to obtain the population inversions: collisions with energetic electrons can excite (and ionize) chemical species. Optical selection rules do not necessarily apply, so that forbidden metastable states may become accessible. Important examples include the argon-ion and nitrogen lasers. It is also possible to use intermolecular energy transfer to generate a population inversion in a species other than the one first excited, and examples include the helium–neon and carbon dioxide lasers. Finally, another class of gas-phase laser is the *excimer* (*exciplex*) laser mentioned in connection with Section 3.9 (see Note 3.27, p. 57).

3.8 Chemiluminescence

Some chemical reactions are accompanied by the emission of light, and the phenomenon is that of *chemiluminescence*. The excitation is *not* thermal; in flames (which show emission characteristic of, for example, the species C_2, CH, and OH), emission intensities may be far higher than those expected from the flame temperature, and the radiation is chemiluminescent. Several natural chemiluminescent phenomena are well known, among them the light of glow-worms and fireflies, the glow of rotting fish, many bacteria, and the cold will-o'-the-wisp.

Chemiluminescence in the firefly system is remarkably efficient, the overall quantum yield for emission approaching unity. The substrate molecule luciferin is oxidized in the presence of the energy-rich phosphate ATP (adenosine triphosphate: see Section 5.3 and Fig. 5.6)

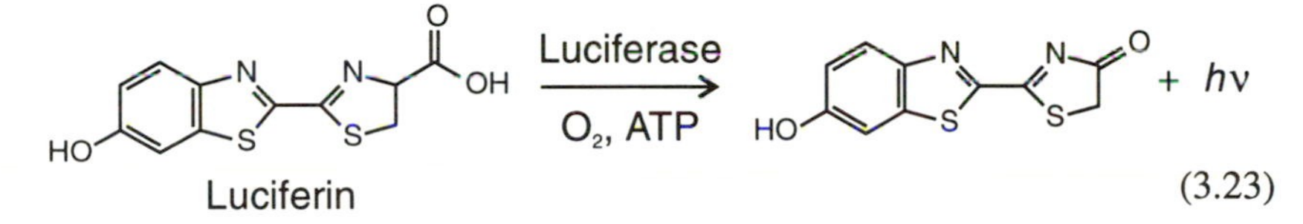

(3.23)

3.25 Formula of CPPO.

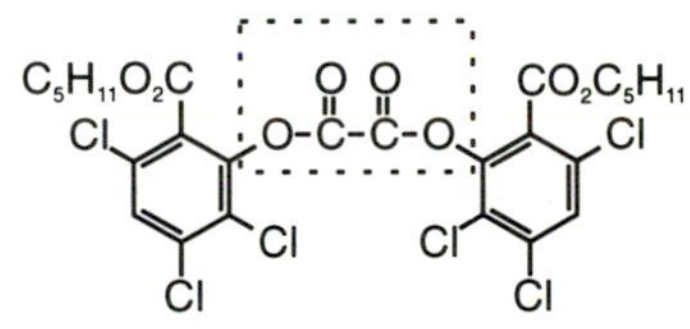

An enzyme *luciferase* is necessary to trigger the process.

One of the most efficient man-made chemiluminescent systems discovered so far is based on aromatic esters of oxalic acid (ethanedioic acid). Hydrogen peroxide decomposes the ester catalytically, and an energy-rich intermediate in the decomposition can transfer energy to a fluorescer present in the mixture. The ester CPPO (bis(carbopentoxy-3,5,6-trichlorophenyl) oxalate) is a typical example of the starting material (see Note 3.25). Quantum yields reach values as high as 0.32, and the spectral characteristics can be selected by suitable choice of the fluorescer. Good fluorescers include 9,10-diphenylanthracene (blue), bis(phenylethynyl)anthracene (green), and rubrene (red). The commercial CYALUME® lightsticks use the oxalate system, the solution of oxalate ester and fluorescer being contained in an outer plastic tube, and the H_2O_2 in an inner breakable glass tube.

Chemiluminescence is of great value in analysis, partly because very low levels of emitted light may be detected (as low as a few photons per second). One standard biochemical assay for ATP uses nature's efficient chemiluminescent materials of firefly luciferin and luciferase, and the oxalate esters may be used to detect picomole quantities of hydrogen peroxide.

3.26 The oxalate system has been proposed for use in emergency lighting systems, particularly in situations where the absence of heat or flame may be vital. Life-jacket markers and ropes using oxalate esters are amongst the many suggested applications. Potential artistic uses of chemiluminescence have been recognized, and necklaces containing re-packaged CYALUME® are a familiar sight at fairgrounds.

Many gas-phase reaction systems are chemiluminescent. Low-pressure gas-phase chemiluminescence can provide detailed information about the nature of the interaction that gives rise to the products, so long as the vibrational and rotational structure is preserved. The study of such emission has found important applications in the investigation of reaction dynamics and the types of potential energy surface that link products with reactants (see Section 2.13).

Two types of process that give rise to excitation are *recombination reactions* and *exchange reactions*, and simple examples are provided by the processes

$$O + NO + M \rightarrow NO_2^* + M \quad (3.24)$$
$$NO + O_3 \rightarrow NO_2^* + O_2 \quad (3.25)$$

Both reactions produce electronically (and vibrationally) excited NO_2, although with differing degrees of excitation so that the emission lies in different parts of the spectrum. Reaction (3.24) leads to visible emission that is apple-green in colour: it is called the 'air afterglow' emission because it was first observed in air that had been subjected to an electric discharge. Reaction (3.25), on the other hand, produces chemiluminescence in the red and near-infrared regions of the spectrum, because the reaction is less exothermic than reaction (3.24). An interesting observation has been made that when NO is deliberately released from a rocket reaching the upper atmosphere, an expanding ring of emission appears which apparently originates from both reactions (3.24) and (3.25), thus demonstrating that both O atoms and O_3 are present at those altitudes in the atmosphere. There are many other chemiluminescent emission processes that occur by day and by night to generate the *atmospheric airglow*, the spectroscopic study of which has yielded much information about composition, chemistry and motions in the upper atmosphere.

3.9 Excimers and exciplexes

The theme of this chapter has been how intramolecular and intermolecular processes involving excited species can affect the emission of light as a result of the exchange of energy. Another important type of interaction is the formation of relatively long-lived molecular complexes of excited species. Studies of emission provide much information about complex formation, and it is appropriate to introduce the concepts here, although it should be recognized that other aspects of photochemistry may be influenced as well.

Many molecules that do not interact significantly in their ground states appear to form reasonably stable complexes when excited. The complexes are called *excimers* or *exciplexes*, words derived from **exci**ted di**mer** and **exci**ted com**plex**. An excimer is produced by the interaction of an excited molecule with a ground-state molecule of the same chemical identity, while an exciplex involves interaction with a chemically different species. Excimers and exciplexes have a fixed and simple stoicheiometric composition, usually 1:1.

Excimers were first recognized by the effect of increasing concentration on the fluorescence of some solutes. The intensity of the normal fluorescence decreases, while a new band at longer wavelengths appears, the intensity of which increases with increasing concentration. Some aromatic hydrocarbons such as pyrene show this behaviour particularly clearly. Pyrene itself has a violet fluorescence when it is in dilute solution, which is replaced by a structureless blue emission at high concentrations. An excimer is formed between an excited singlet pyrene and ground-state pyrene, and it is the excimer that radiates. If pyrene is represented as P, the excitation scheme is

$$P(S_0) + h\nu \rightarrow P(S_1) \quad \text{absorption} \quad (3.26)$$

$$P(S_1) \rightarrow P(S_0) + h\nu \quad \text{normal fluorescence} \quad (3.27)$$

$$P(S_1) + P(S_0) \rightarrow P(S_1){:}P(S_0) \quad \text{excimer formation} \quad (3.28)$$

$$P(S_1){:}P(S_0) \rightarrow P(S_0) + P(S_0) + h\nu \quad \text{excimer fluorescence} \quad (3.29)$$

The emission from the excimer is initially to a $P(S_0){:}P(S_0)$ pair possessing the same geometry as the excimer, but, since there is no attraction between the ground-state molecules, it has the same energy as isolated $P(S_0)$ molecules, and immediately separates. The excimer, on the other hand, *is* stabilized with respect to an isolated $P(S_1)$ molecule, so that the emission from the excimer lies at longer wavelengths than the normal fluorescence. Excimer emission is structureless, because the lower state of the transition is essentially a continuum resulting from the repulsion of the ground-state molecules.

Exciplex emission is seen in solutions of mixed solutes. For example, fluorescence of anthracene is quenched by diethylaniline, and a new and structureless emission is observed at longer wavelengths. The emission is not sensitized fluorescence of diethylaniline, but rather emission from the excited singlet anthracene–diethylaniline complex.

Excimer formation occurs with non-aromatic molecules as well as with aromatic ones, and triplet excimers also exist, giving rise to excimer phosphorescence. The excimer of pyrene is, however, particularly strongly bonded, thermodynamic measurements indicating a bond strength of around $40\,\text{kJ mol}^{-1}$. Experimental results suggest that the dimer is quite rigid, and it is thought that the aromatic hydrocarbons adopt a sandwich structure with a separation between the planes of about 0.33 nm.

The nature of bonding in excimers and exciplexes is clearly dependent on the presence of electronic excitation. Part of the stabilization arises from the promotion of an electron from a filled orbital that would be antibonding in a ground-state dimer or complex to an unfilled bonding orbital in the excited pair. Charge-transfer stabilization is also important in excimers, and especially in exciplexes. The promotion of an electron to a higher energy level obviously makes an excited molecule potentially a better donor than the ground-state molecule, but the lower-energy orbital from which the electron was promoted can also now receive another electron, thus leading to increased acceptor properties (see Section 2.2). Electrostatic attraction between M^+ and N^- can thus arise in an excited pair $(MN)^*$, the direction of electron transfer depending on the particular chemical species involved.

3.27 The phenomena described here provide the route to another way of achieving population inversion in an important class of laser (see Section 3.7), the *excimer laser* (although exciplexes are really involved). The lifetime of the ground-state exciplex or excimer is around one vibrational period and its population is negligible. Formation of the excited-state complex necessarily gives it a population greater than that of the hypothetical ground state, and laser operation is possible. Important practical examples include noble gas–halogen systems such as ArF and XeCl.

4 Photochemical kinetics

4.1 Reaction rates and photochemistry

In this chapter, we shall show how an analysis of information about the rates of reactions can lead to a better and more quantitative understanding of photochemical processes. *Reaction kinetics* is the part of chemistry that deals with reaction rates, and photochemical kinetics affords valuable insights into the mechanisms of photochemical reactions and even into the detailed nature of individual elementary reaction steps. The kinetic approach is also a valuable adjunct to studies of absorption spectra, fluorescence, and many other optical and photochemical phenomena. Its use has been implicit in much of the discussion of earlier chapters, and it is the intention now to give a more specific account of the relation between kinetics and photochemistry.

4.2 Multistep reaction schemes

One feature of many photochemical processes is that secondary steps follow the primary event that follows the initial absorption of light (see p. 10 and Note 1.20). Kinetic tools are of particular value in unravelling the complexities of such *multistep* systems.

Stationary-state treatments

A simple example will show the kinetic approach to the elucidation of reaction mechanisms and indicate how rate constants can be extracted from experimental data. The photolysis of mixtures of ozone and oxygen by red light might be expected to proceed via the mechanism

$$O_3 + h\nu \xrightarrow{\phi_1} O + O_2 \tag{4.1}$$

$$O + O_3 \xrightarrow{k_2} O_2 + O_2 \tag{4.2}$$

$$O + O_2 + M \xrightarrow{k_3} O_3 + M \tag{4.3}$$

The appearance of M in eqn (4.3) is explained in Note 4.1. It is assumed that the quantum yield for the primary process, ϕ_1, is unity, in accordance with experimental observation. The rate of the production of O atoms in reaction (4.1) is then just equal to I_{abs}, the intensity of light absorbed [Note 4.2]. Thus, the rate equations for the formation of O and loss of O_3 are

$$\frac{d[O]}{dt} = I_{abs} - k_2[O][O_3] - k_3[O][O_2][M] \tag{4.4}$$

$$-\frac{d[O_3]}{dt} = I_{abs} + k_2[O][O_3] - k_3[O][O_2][M] \tag{4.5}$$

4.1 The 'third-body', M, is necessary in reaction (4.3), as in many other recombination reactions, to stabilize the energy-rich O_3 molecule initially formed; the molecule would otherwise fall apart again.

4.2 The rate of production of O atoms in reaction (4.1) is $\phi_1 I_{abs}$; ϕ_1 is approximately unity, so that the production rate is simply I_{abs} (see beginning of Section 4.3 for an explanation).

Multistep reaction schemes are interpreted kinetically by writing down the differential equations, such as eqn (4.4.) or (4.5), for all the species of interest, including the intermediates. Solution of these equations then allows the variation of the concentration of each of the species with time to be predicted. Analytical (i.e. algebraic) solution of the many simultaneous differential equations is rarely possible, but numerical solutions using computers are widely used. Such methods do not, however, afford much insight into the underlying chemistry of the system. For some highly reactive intermediates [Note 4.3], the *stationary (or steady) state hypothesis* (SSH) provides a simplification that will permit algebraic solution of the kinetic equations. The idea behind the SSH is that an intermediate, X, can reach a concentration that is very nearly constant with time. So long as the rate of loss of X increases with increased concentration of X, after the reaction is started [X] will then increase until the rate of its loss is equal to that of its formation. A *steady state* for [X] has been reached, and d[X]/d*t* is very nearly zero.

4.3 The reactive intermediates encountered in photochemistry, to which the text refers, are atoms, radicals, electronically and vibrationally excited species, and sometimes ions.

In applying the SSH to our example of ozone photolysis, we set the differential in eqn (4.4) to zero, since atomic oxygen is a highly reactive intermediate. To simplify the appearance of the equations, let us rewrite eqn (4.4) in the form

$$\frac{d[O]}{dt} = I_{abs} - k'[O] \tag{4.6}$$

where $k' = k_2[O_3] + k_3[O_2][M]$. We proceed now on the assumption that the SSH hypothesis holds for this system, and return later to the question of its validity. Setting eqn (4.6) to zero yields the value for the steady-state concentration of oxygen

$$[O]_{ss} = I_{abs}/k' \tag{4.7}$$

This value for the atomic oxygen concentration can then be used for substitution in eqn (4.5) to obtain the rate of ozone loss. Performing this substitution leads to the result for the rate of photolysis

$$\text{rate} = -\frac{d[O_3]}{dt} = \frac{2I_{abs}k_2[O_3]}{k_2[O_3] + k_3[O_2][M]} \tag{4.8}$$

Inversion of eqn (4.8) yields the result

$$\frac{1}{\text{rate}} = \frac{1}{2I_{abs}}\left(1 + \frac{k_3[O_2][M]}{k_2[O_3]}\right) \tag{4.9}$$

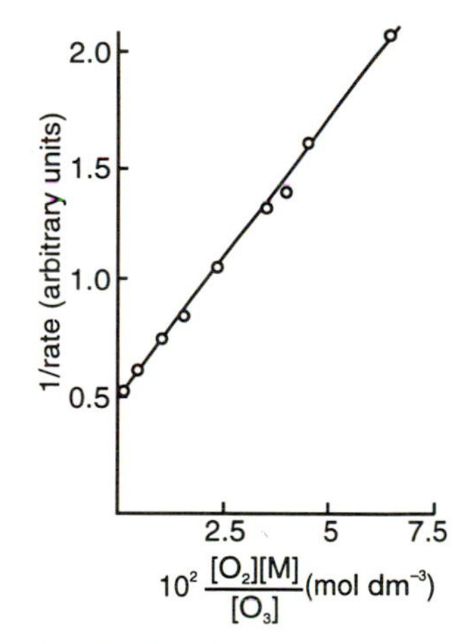

Fig. 4.1 Plot of 1/rate against $[O_2][M]/[O_3]$ for photolysis of ozone by red light.

Figure 4.1 shows data for some real experiments plotted in the form of eqn (4.9). The graph is essentially linear, which provides strong, but circumstantial, evidence that the hypothetical mechanism is correct. Information about the rate constants k_2 and k_3 can also be obtained from the graph. The units used to measure the rate of reaction are arbitrary, and the absolute intensity of light absorbed need not be measured: the ratio of the slope to intercept is simply k_3/k_2, because the scaling factors for rate and intensity terms are the same and therefore cancel [Note 4.4]. This ratio fits well with independent measurements of the individual values for k_2 and k_3 (see p. 61).

4.4 From eqn (4.9), it is evident that a graph of 1/rate versus $[O_2][M]/[O_3]$ should have an intercept of $1/(2I_{abs})$ and a slope of $k_3/(2I_{abs}k_2)$. It therefore follows that the ratio of slope to intercept is k_3/k_2, the two I_{abs} terms cancelling each other out (see also p. 64).

Validity of the steady-state assumption

We must now return to the question of the applicability of the SSH to the ozone photolysis system. The problem is to know if the concentration of O calculated using the SSH bears any relation to actual concentrations. Our example has been chosen because it can also be solved analytically. So long as the extent of photolysis is small, I_{abs}, $[O_2]$, $[O_3]$ and [M] can all be taken as nearly independent of time, and eqn (4.6) can be integrated to yield

$$[O] = \frac{I_{abs}}{k'}[1 - \exp(-k't)] \tag{4.10}$$

where t is the time for which the system has been illuminated. In a typical experiment, k' might turn out to be 400 s^{-1}. Dividing eqn (4.10) by (4.7) gives the ratio $[O]/[O]_{ss}$ as $[1-\exp(-k't)]$; values of this ratio are given for different times in Note 4.5. It can be seen that, so long as t is greater than about 0.01s, the true concentration approaches the one calculated by the SSH to within two percent, and the SSH approximation is probably good enough. This conclusion suggests that the steady-state treatment would be perfectly adequate in dealing with an experiment in which ozone was being photolysed slowly over a period of minutes or hours. We should recognize, however, that the treatment would not be satisfactory if the concentrations of reactants were several orders of magnitude smaller than those postulated, or if measurements on a smaller time scale were required. In principle, the applicability of the SSH should be established for any particular kinetic system before it is used. The reason why the SSH can be applied to highly reactive intermediates should now be clear: the more reactive the intermediate, the greater the value of the rate constant for loss (for example k' in eqn (4.10)), and the shorter the time taken for the steady state to be approached sufficiently closely.

4.5 Approach to steady state (for $k'=400\,s^{-1}$)

t/s	$[O]/[O]_{ss}$
10^{-3}	0.33
10^{-2}	0.98
1	~1

Time-resolved experiments

One feature of steady-state experiments, which is highlighted by the example of ozone photolysis, is that only *ratios* of rate constants can be obtained from the kinetic analysis. This result is a consequence of the steady state being established because of competition between production and loss processes. Although much valuable information can be extracted, especially if one of the rate constants involved can be determined absolutely in some other way, the limitation does seriously restrict the applicability of steady-state experiments to kinetic investigations. The alternative approach is to study the reaction system in a *time-resolved* manner, using *non-stationary* conditions. For highly reactive intermediates, very little time is taken to reach the steady state; the non-stationary experiments have therefore depended upon the development of suitable techniques with very fast time resolution. In photochemistry, one of the most powerful and important of these has been *flash photolysis* (p. 17). A short-duration flash of light from a discharge tube or laser is used to initiate the photochemical reaction, and the transient changes in concentration of reactants, products, or, most frequently, the intermediates themselves are monitored as a function of time. Successive developments have brought the time-scale of flash photolysis down from 10^{-3} s in early experiments (a major advance at the time) to as little as 10^{-15} s in some specialized current experiments. One important aspect of the flash-photolysis technique is that it has permitted positive identification, by direct spectroscopic methods, of reaction intermediates whose existence had hitherto been hypothetical.

To explain how flash-photolysis experiments may be used to determine rate constants, let us revert to the example of ozone photolysis. Consider what would happen if the ozone were exposed to a flash of light rather than to steady illumination. Immediately after the flash, some of the ozone would have been dissociated to form atomic oxygen, but no further O can be formed. The losses of atomic oxygen in reactions (4.2) and (4.3) will continue, and eqn (4.6) simplifies, in the absence of light and therefore of the I_{abs} term, to

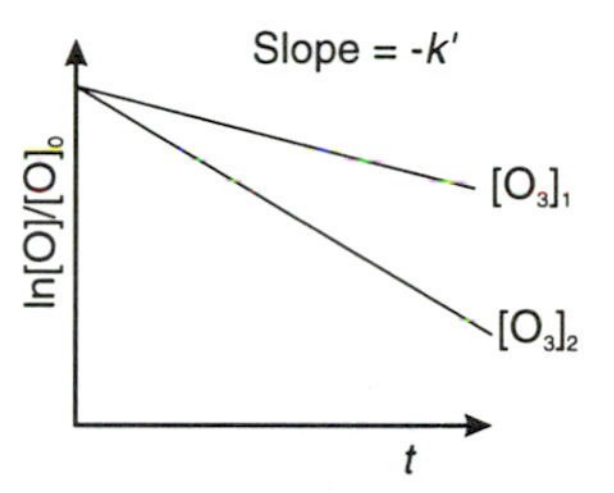

Fig. 4.2 Plot of $\ln[O]/[O]_0$ as a function of t, showing the decay of atomic oxygen for two initial ozone concentrations.

$$\frac{d[O]}{dt} = -k'[O] \qquad (4.11)$$

So long as the extent of the secondary reactions is small, $[O_3]$, $[O_2]$, and thus k' can be assumed to be roughly constant, and the differential equations can be integrated to

$$\begin{aligned}[O] &= [O]_0\exp(-k't) \\ &= [O]_0\exp\{-(k_2[O_3] + k_3[O_2][M])t\} \qquad (4.12)\end{aligned}$$

where $[O]_0$ is the concentration of atomic oxygen immediately after the flash. Atomic oxygen concentrations can be measured in flash-photolysis experiments by direct techniques, including absorption and fluorescence methods. Equation (4.12) indicates that if the (natural) logarithm of $[O]/[O]_0$ is plotted as a function of t, a straight line will result. Figure 4.2 shows the types of decay that might be expected for two different initial ozone concentrations. The slopes of these lines are equal to $-k'$, and a further plot of k' against $[O_3]$ (Fig. 4.3) produces a straight line of slope k_2 and intercept $k_3[O_2][M]$. Since $[O_2][M]$ is known, both k_2 and k_3 can thus be calculated. The significant feature of eqn (4.12), in the context of this section, is that the two rate constants are separated out in a sum, rather than combined in a ratio as in the steady-state expression (4.9).

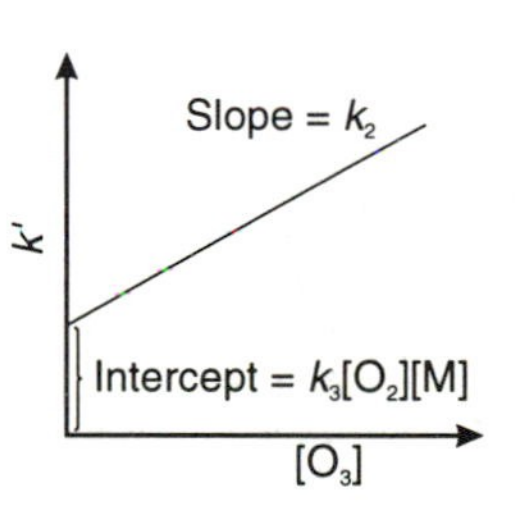

Fig. 4.3 Plot of k' as a function of $[O_3]$.

4.3 Quantum yields and rates

The concept of the *quantum yield* of a photochemical process was introduced in Section 1.8. The *primary* yield, ϕ, gives an indication of the likelihood of the decay of a species via a particular primary channel, such as dissociation or fluorescence, amongst all the various different channels available (see

Fig. 1.5). Since each photon absorbed excites one molecule (the Stark–Einstein Law, p. 10), ϕ is the number of molecules undergoing the process of interest divided by the numbers of photons absorbed [Note 4.6]. Quantum yields may also be expressed in terms of *rates* of reaction rather than in terms of absolute numbers. An *intensity* of radiation, I, refers to an energy per unit time, and it is frequently convenient to express the *absorbed intensity*, I_{abs}, as the energy absorbed in unit time by unit volume: it is then in the same form as a rate expressed in concentration units, with energies (numbers of photons) replacing concentrations (numbers of molecules). We can therefore express ϕ in two ways

4.6 Of course, any one excited molecule has to decay in one particular manner. However, ordinary sized samples contain huge numbers of molecules, and ϕ represents the fraction of these decaying by the chosen route.

$$\phi = \frac{\text{number of chemical events}}{\text{number of photons absorbed}} \tag{4.13}$$

$$\phi = \frac{\text{number of chemical events/unit time}}{\text{number of photons absorbed/unit time}}$$

$$= \frac{\text{rate of chemical process}}{\text{intensity of light absorbed}} = \frac{d[X]/dt}{I_{abs}} \tag{4.14}$$

where $d[X]/dt$ is the rate of the particular channel under discussion.

The second part of the expression in eqn (4.14) is the most convenient way of handling quantum yields in kinetic expressions, because the rate of the process, $d[X]/dt$, is simply equal to ϕI_{abs}. This is the form of equation that we have already used for the rate of the primary process earlier in this chapter (see Section 4.2, with $\phi = 1$ for reaction (4.1)). Note that, if the particular channel is not specified, the rate of reaction involved is *the rate of disappearance of reactant*.

The treatment just presented also provides a justification for the way that rates have been used to calculate quantum yields, as for example in connection with fluorescence on p. 42. The procedure has been to take the ratio of the rate of the process of interest (eg, fluorescence) to the sum of the rates of all processes occurring (eg, fluorescence, quenching, and ISC). Since the sum, ϕ_{total}, of the quantum yields for the individual processes ϕ_i, is unity (see eqn (3.1)), examination of the equation

$$\phi_i = \frac{d[X_i]/dt}{I_{abs}}; \qquad \phi_{total} = 1 = \sum \frac{d[X_i]/dt}{I_{abs}} \tag{4.15}$$

makes it evident that

$$\phi_i = \frac{d[X_i]/dt}{\sum d[X_i]/dt} \tag{4.16}$$

The occurrence of secondary reactions can make the *overall quantum yield*, Φ, for a photochemical reaction exceed the primary quantum yield by a factor that depends on the number of reaction steps that follow the initiating, primary one [Note 1.20]. Thus, if ϕ is unity, Φ can be greater than unity, and can be much bigger if a chain reaction follows the primary step. A quantity often discussed in connection with thermal chain reactions is the *kinetic chain length*, ν. This quantity is the number of hypothetical 'links' in the chain (i.e. the number of times the reactions occur), and is thus the total number of reactant molecules consumed for each chain started, or, in rate terms [Note 4.7],

$$\nu = \frac{\text{rate of reactant consumption}}{\text{rate of initiation}} \tag{4.17}$$

Since, in a photochemical reaction, the rate of formation of intermediates, and thus the rate of initiation of the secondary reactions, is ϕI_{abs}, and the overall rate of reaction is ΦI_{abs}, it is evident that $\nu = \Phi/\phi$.

An example will illustrate how kinetic expressions for rates and quantum yields are related very simply to each other. Consider the example of ozone photolysis by red radiation with which we started our discussion of steady-state kinetics. Equation (4.8) gives the rate of reaction derived using the stationary state hypothesis. The overall quantum yield, Φ, for this process is just $-d[O_3]/dt$ divided by I_{abs}, and is thus given by

$$\Phi = \frac{-d[O_3]/dt}{I_{abs}} = \frac{2k_2[O_3]}{k_2[O_3] + k_3[O_2][M]} \tag{4.18}$$

It is worth noting that this equation predicts a value of Φ approaching two when $[O_3]$ is large, and this is the value found by experiment.

4.7 Equation (4.17) is stated in the same form as eqn (4.14). The number of molecules consumed per unit time is the rate of reactant consumption and the number of chains started per unit time is the rate of initiation.

4.4 Kinetics and quantum yields of emission processes

Steady-state quenching: the Stern–Volmer relation

In Chapter 3 (p. 42; see also Fig. 3.2), we introduced the Stern–Volmer relation that describes the decrease in intensity of fluorescence brought about by addition of a quencher. The subject was presented in terms of the competing reactions, using the approach embodied in eqn (4.16) that we have just derived. It is instructive and useful to re-examine the simple quenching process explicitly in terms of intensities and rates of reactions. The photochemical scheme, in its simplest form, consists of the reactions

$$A + h\nu \rightarrow A^* \tag{4.19}$$

$$A^* \rightarrow A + h\nu \tag{4.20}$$

$$A^* + M \rightarrow A + M \tag{4.21}$$

which are, respectively, excitation by absorption, fluorescent emission, and quenching. With the definitions of the rate constants given in Note 4.8 (see also p. 42), we may write the rate equation for A^* as

4.8 The rates for the reactions given in the text are
(4.19) I_{abs}
(4.20) $A[A^*]$
(4.21) $k_q[A^*][M]$

$$\frac{d[A^*]}{dt} = I_{abs} - A[A^*] - k_q[A^*][M] \tag{4.22}$$

This equation is the exact equivalent of eqn (4.4). In the steady state, this equation may be set equal to zero, so that

$$[A^*] = I_{abs}/(A + k_q[M]) \tag{4.23}$$

The fluorescence emission intensity, I_f, is the rate of process (4.20)

$$I_f = A[A^*] = I_{abs}A/(A + k_q[M]) \tag{4.24}$$

On inversion, this equation becomes

$$\frac{1}{I_f} = \frac{1}{I_{abs}}\left(1 + \frac{k_q[M]}{A}\right) \tag{4.25}$$

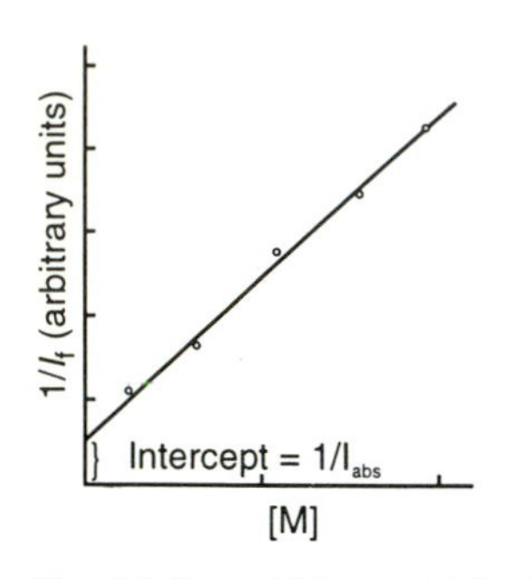

Fig. 4.4 Stern–Volmer plot for the quenching of fluorescence.

Because $I_f/I_{abs} = \phi_f$ by definition, so that I_{abs}/I_f is equal to $1/\phi_f$, this equation is exactly the same as eqn (3.5) although it has been derived in a slightly different way. Figure 4.4 shows the same data as were given in Fig. 3.2, but now plotted out in intensity form. The intercept is $1/I_{abs}$ and the slope is $(1/I_{abs}) \times (k_q/A)$ so that the ratio of slope to intercept is the ratio k_q/A discussed in connection with eqn (3.5). Measurement of *absolute* intensities is difficult (see pages 14–15), and it is usual to measure both I_{abs} and I_f in arbitrary units. Because the slope is divided by the intercept in the present calculation, the scaling factor for the units cancels out. The ratio is thus independent of whatever the true calibration factor may be.

Lifetimes in emission studies

As explained on pages 60–61, time-resolved photochemical experiments can potentially yield much more detailed information than steady-state ones. The advantages of time-resolved measurements are particularly evident in the measurement of emission lifetimes of fluorescence and phosphorescence.

In order to explain the meaning of the expression *lifetime* in kinetic terms, it is useful first to describe the kind of experiment that is behind all the studies envisaged here. A sample of luminescent material is irradiated with a brief flash of light, as in flash photolysis. The flash might typically last 10^{-10} s in a present-day experiment, while the decay of emission might be measurable over 10^{-7} s or more. For the present purposes, therefore, it is perfectly reasonable to consider that there is no exciting radiation present at all after the flash, during which time the intensity measurements are made. For a species A excited to A* in a pulsed form of reaction (4.19) and decaying in reactions (4.20) and (4.21), the differential kinetic equation is closely similar to eqn (4.11), and is

$$\frac{d[A^*]}{dt} = -(A[A^*] + k_q[A^*][M]) = -k'[A^*] \tag{4.26}$$

As in all the previous examples, the losses of A* are effectively all first order (processes such as reaction (4.21) are pseudo-first order, because [M] does not vary with time), so that it is possible to simplify the equations by writing a combined first-order rate constant, k'. The integrated form of eqn (4.26) is the same in form as eqn (4.12), and, since $I_f = A[A^*]$, it follows that

$$\begin{aligned} I_f(t) &= I_f(0)\exp(-k't) \\ &= I_f(0)\exp\{-(A + k_q[M])t\} \end{aligned} \tag{4.27}$$

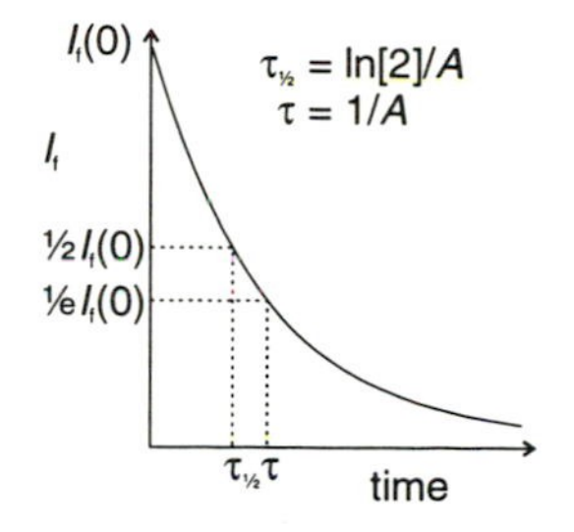

Fig. 4.5 Plot of I_f as a function of t showing τ and $\tau_{½}$ for the exponential decay implied by eqn (4.27).

where $I_f(0)$ and $I_f(t)$ are the emission intensities at zero time and at the measurement time, t, respectively. The exponential decay curve implied by this equation is represented in Fig. 4.5. The markings in the figure show both the ordinary half-life, which is the time taken for I_f to decay to half of its initial intensity, and the rather simpler lifetime, τ, which is the time taken for I_f to decay to 1/e of its initial value; the simplicity is mathematical, since τ is equal to $1/A$ [Note 1.9].

It is usual to plot the decay of intensity with a logarithmic scale for the ordinate (y-axis), as illustrated in Fig. 4.6; the slopes are just $-k' = -(A + k_q[M])$. Results are shown for three different concentrations of quencher [M], and illustrate clearly the dependence of k' on [M]. The final step in extracting the individual rate constants is to plot the values of k' from these slopes against [M] in a further graph, shown here as Fig. 4.7. The slope of this graph is k_q and the intercept is A.

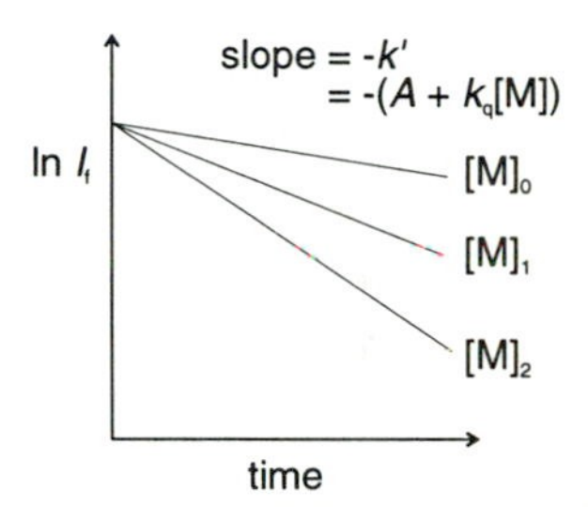

Fig. 4.6 Plot of $\ln I_f$ as a function of t.

Lifetimes in more complex cases: radiationless processes

The discussion of Chapter 3 will have shown how intramolecular radiationless energy transfer processes can add to the emission and quenching losses in real cases of fluorescence and phosphorescence. For example, another ISC step, $S_1 \rightsquigarrow T_1$

$$A^*(S_1) \rightsquigarrow A(T_1) \tag{4.28}$$

can add to the three simplest reactions (4.19) to (4.21). One consequence of ISC is that the effective lifetime is shorter than the 'natural' radiative lifetime ($1/A$), even though no collisions are occurring. The way in which the rate constant for the process, say k_{ISC}, is obtained provides an interesting example of how both steady-state and time-resolved studies may be combined. In order to make the example that follows reasonably simple, it will be assumed that bimolecular quenching, reaction (4.21), can be neglected. If it cannot, then the procedures described in the previous section, and the extrapolation to zero [M] embodied in Fig. 4.7, can be used to remove the effects of quenching. Equation (4.26) now becomes

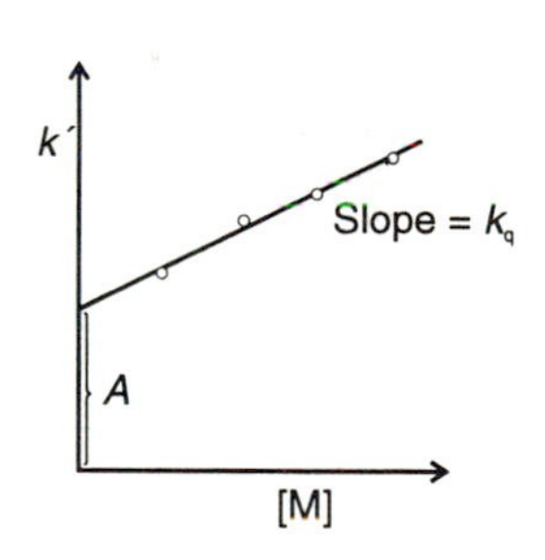

Fig. 4.7 Plot of k' as a function of [M].

$$\frac{d[A^*]}{dt} = -(A[A^*] + k_{ISC}[A^*]) = -k'[A^*] \tag{4.29}$$

This equation is the same as eqn (4.26), but with $k_q[M]$ replaced by k_{ISC}.

The problem is that the two terms making up k' are just additive $(A + k_{ISC})$. Time-resolved experiments will provide this sum, but cannot, as they stand, resolve the sum into its two components. One solution is to make steady-state intensity measurements as well. The analogue of eqn (4.25) for this situation is

$$\frac{1}{I_f} = \frac{1}{I_{abs}}\left(1 + \frac{k_{ISC}}{A}\right) \tag{4.30}$$

so that measurement of both I_{abs} and I_f will allow the value of k_{ISC}/A to be obtained. The sum of A and k_{ISC} is known from the time-resolved experiments, and the ratio from the steady-state ones. It is now a simple matter to calculate the individual values [Note 4.9].

4.9 The individual values come from the sum and the ratio by simple manipulation

If $a = A + k_{ISC}$

and $b = k_{ISC}/A$

$k_{ISC} = A.b$

∴ $a = A + A.b$

$= A(1 + b)$

∴ $A = a/(1 + b)$

and $k_{ISC} = a.b/(1 + b)$

Absolute emission intensities

The steady-state calculations just described require determination of both I_{abs} and I_f, the measurements corresponding to a determination of the absolute quantum yield for fluorescence. Strictly speaking, it is not necessary to measure the intensities on an absolute scale, because only their ratio is used (see pages 60 and 64). They must, however, be measured on the same relative scale, so that a link has to be made between the absorbed intensity, which is likely to be high, and the emitted intensity, which can be weak. Absolute intensity measurements were described in Section 1.10. The same principles apply to the determination of emission intensities, with a requirement ultimately to measure absolute energies and convert them to intensities through Planck's Law. The problem lies in the low energies often encountered. There are a number of emission systems (fluorescent and chemiluminescent) for which the absolute yields have been measured and which can be used in the same manner as chemical actinometers, possibly in order to calibrate photomultipliers or other detectors of radiation. These studies are demanding, and it is fortunate that many of the important measurements on emission systems do not, in fact, require absolute measurements to be used.

5 Photochemistry in nature

5.1 Introduction

Earlier chapters of this book have been concerned with the fundamental processes of photochemistry. In this and the next chapter, we look at some of the many ways in which photochemistry has an impact on our lives. Natural photochemical phenomena have contributed to the evolution of life as we know it, and permit its continued existence on Earth.

5.2 Atmospheric photochemistry

Origin and evolution of the atmosphere

Photochemical reactions have performed a determining role in the evolution of the atmosphere and of life on Earth. Our understanding of primary photochemical processes permits reasonable speculation about the history of the atmosphere; the 'investigation' of the Earth's palaeoatmosphere (fossil atmosphere) has in turn suggested solutions to several 'puzzles' concerning the Earth's geology. The forms and ecology of life that were viable at any time in the past were directly dependent on the constitution of the atmosphere at that period; we shall see that, on the other hand, processes involving living organisms exert a major influence on atmospheric composition. It is this interrelation between atmospheric and biological evolution that makes the study of the Earth's palaeoatmosphere, and comparison with the present-day atmospheres of other planets, particularly rewarding.

Inorganic photochemical processes are unable to explain the levels of O_2 found in our atmosphere. How could our atmosphere come to have more than 20% of oxygen in it, while our nearest neighbours, Venus and Mars, have less than 0.1%? Earth possesses an atmosphere that for hundreds of millions of years appears to have been disregarding the laws of physics and chemistry. Minor oxidizable constituents of our atmosphere, such as methane, ammonia, hydrogen, carbon monoxide, and nitrous oxide, survive in the presence of large concentrations of oxygen, although thermodynamic considerations would suggest virtually complete oxidation of these components. Earth's peculiar behaviour is a consequence of the existence of life on the planet. Biological processes, acting together with physical and chemical change, determine the composition of our atmosphere. Conversely, our unique atmosphere seems essential for the support of life in its many forms. Oxygen, the unexpected gas of our atmosphere, is almost entirely the result of biological activity. Not only does biology provide the atmospheric oxidant, but it also continually provides the oxidizable minor gas 'fuels'. Biological processes are thus responsible for the thermodynamic disequilibrium of our atmosphere.

5.1 Much evidence shows that Earth was without a primordial atmosphere. For example, the abundances of the noble gases in the contemporary atmosphere lie between 10^{-10} and 10^{-6} of their cosmic abundances. It has been shown that the quantities of gases liberated as a result of volcanic activity, and from slow decay of solid radioactive elements, are sufficient to account for our atmosphere. However, *oxygen is not released from volcanic effluents*, and the primitive atmosphere must have contained N_2, CO_2, and H_2O as its most important constituents, together with traces of reducing gases such as H_2 and CO.

Photosynthesis is the only known process that can yield O_2 in its present abundance. Photosynthesis is the subject of Section 5.3, and for the time being we need only note that the process involves consumption of carbon dioxide and water, and the concomitant *liberation of oxygen* [Note 5.2]. In the present atmosphere, all O_2 passes through the photosynthetic process in a few thousand years, a period extremely short by geological standards, and photosynthesis is clearly an efficient source of O_2. Figure 5.1 summarizes one view of the build-up of oxygen from pre-biological to present-day levels as large-scale photosynthesis became established.

5.2 The simplified equation for photosynthesis is

$$n\mathrm{CO_2} + n\mathrm{H_2O} \xrightarrow{h\nu} (\mathrm{CH_2O})_n + n\mathrm{O_2}$$

and is met again as eqn (5.16). Almost all the oxygen in the Earth's atmosphere is produced by photosynthesis. Figure 5.1 shows how oxygen concentrations might have evolved since life first appeared on the planet some 3.6×10^9 or more years ago. The markers on the oxygen curve show fossil and geological evidence for the particular concentrations. The curve for ozone concentrations is obtained by a model calculation for each of the oxygen concentrations. The results suggest that sufficient protection from solar UV would have been afforded by ozone for life to emerge onto dry land around 550×10^6 years ago.

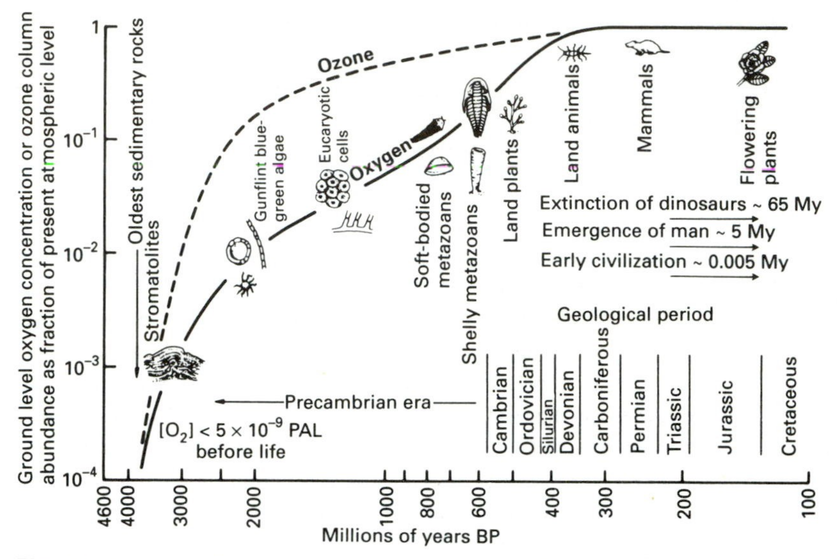

Fig. 5.1 Evolution of oxygen, ozone, and life on Earth.

Pre-biological oxygen concentrations in the atmosphere are important in two ways. Organic molecules are susceptible to thermal oxidation and photooxidation, and are unlikely to have accumulated in large concentrations in an oxidizing atmosphere. The low pre-biological oxygen concentrations thus seem essential to the development of the organic precursors of life. Living organisms can develop mechanisms that protect against oxidative degradation, but they are still photochemically sensitive to short-wavelength ultraviolet radiation. Macromolecules, such as the proteins and nucleic acids that are characteristic of living cells, are damaged by ultraviolet radiation of wavelength shorter than about 290 nm. In our atmosphere, the oxygen itself is able to filter out solar ultraviolet radiation with wavelengths shorter than about 230 nm. For wavelengths between 230 and 290 nm, however, some other protection must be afforded. Luckily, there is a suitable absorbing species in our atmosphere, so that organisms can live on dry land more or less exposed to the filtered Sun's rays. The species is ozone, O_3, derived photochemically from O_2 in the atmosphere (see p. 69). The amount of ozone in the atmosphere, and its altitude distribution, will depend on the concentration of the precursor oxygen, and will therefore have altered markedly as the atmosphere evolved. Concentrations of ozone are also controlled by the rates of loss processes for the molecule. Loss is regulated by catalytic cycles involving other trace gases of the atmosphere, such as the oxides of nitrogen, that are themselves at least partly of biological origin (see p. 70). We have already noted that the oxygen in the Earth's atmosphere comes largely from biological sources. Now we see that the ozone, needed as a filter to protect life, has its concentration determined not only by the biologically generated oxygen required for its production, but also by the biologically generated trace gases that play a part in its destruction.

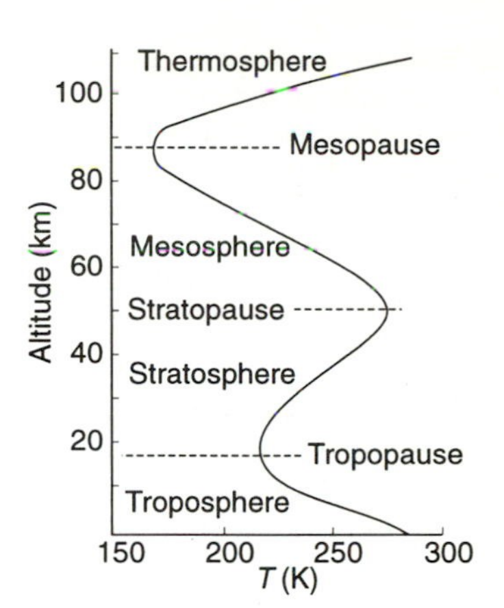

Fig. 5.2 Temperature–altitude profile for the Earth (at a latitude of 40°N during June).

The stratosphere

Simple thermodynamic arguments suggest that temperature should decrease with altitude in the atmosphere. In the Earth's atmosphere, the temperature drops by about 6.5 K for every kilometre of height increase, for roughly the first 15–20 km above the surface, but above this height the temperature begins to increase again. The reversal in the temperature trend constitutes a *temperature inversion*: it results mainly from solar photodissociation of ozone, and the subsequent exothermic chemical reactions that we shall discuss shortly. The lowest region of the atmosphere has colder air lying on top of warmer air, so that convection can lead to rapid vertical mixing: the region is named the *troposphere* after the Greek for 'turning'. In the second region, the warmer air lying on top of colder air results in great vertical stability, and the region is named the *stratosphere* after the Latin for 'layered'. The troposphere and the stratosphere are divided by the *tropopause*. Figure 5.2 shows schematically the temperature–altitude profile in the atmosphere.

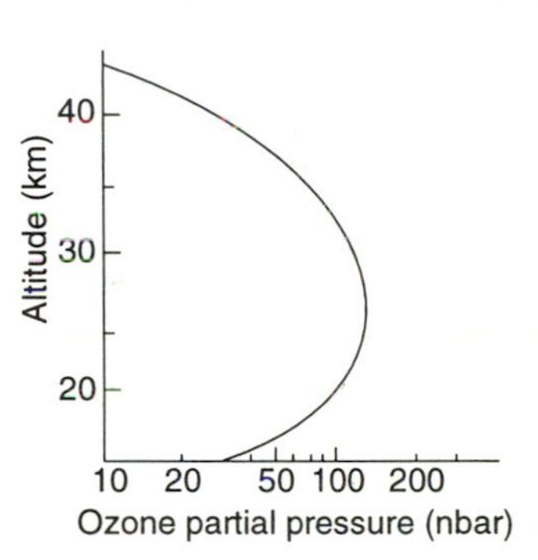

Fig. 5.3 Variation of atmospheric ozone concentration with altitude, as measured with instruments carried on a balloon.

Figure 5.3 shows the results of a typical balloon investigation of atmospheric ozone concentration. The concentration reaches a maximum at an altitude of around 27 km which is quite sharply peaked (note that the concentration is on a logarithmic scale), and atmospheric ozone is frequently described as consisting of a 'layer' in the stratosphere, centred on 25–30 km.

The basic processes that establish the ozone layer were established by Chapman as long ago as 1930. The important 'oxygen-only' reactions are

$$O_2 + h\nu \rightarrow O + O \quad \text{for } \lambda < 242.4\ \text{nm} \qquad (5.1)$$
$$O_3 + h\nu \rightarrow O_2 + O \quad \text{for } \lambda < 1180\ \text{nm} \qquad (5.2)$$
$$O + O_2 + M \rightarrow O_3 + M \qquad (5.3)$$
$$O + O_3 \rightarrow O_2 + O_2 \qquad (5.4)$$

The simple four-reaction scheme predicts the layer structure found in the atmosphere. At high altitudes, there is much short-wavelength ultraviolet radiation capable of dissociating molecular oxygen, but relatively little of the O_2 itself. Low in the atmosphere, there is plenty of O_2, but short-wavelength radiation is absent, because it has already been filtered out by the O_2 and the O_3 lying above. Solar ultraviolet energy absorbed by the O_2 and O_3 is ultimately liberated as heat, in part through the exothermic reactions (5.3) and (5.4). This heating gives rise to the stratospheric temperature inversion.

5.3 Because reactions (5.2) and (5.3) can interconvert atomic oxygen and ozone, O and O_3 are identified as the family of 'odd oxygen'. Reaction (5.1) creates two odd oxygens, and reaction (5.4) destroys two, while reactions (5.2) and (5.3) themselves obviously leave the odd-oxygen concentration unaltered, although they do affect the ratio of [O] to [O_3]. After sunset, atomic oxygen concentrations fall rapidly at altitudes below about 40 km, since the source reactions (5.1) and (5.2) are cut off, but the sink processes (5.3) and (5.4) remain.

Proper calculation of the ozone profile, using rate parameters for the reactions determined in the laboratory, shows that the predicted profile has the same general shape as the measured one. However, the calculated absolute concentrations are all higher, by a factor of up to four or five, than the true atmospheric ones. The problem arises because the loss process, reaction (5.4), is too slow at stratospheric temperatures (say 220–270 K) to balance the production of ozone at the correct concentration. It is now well established that reaction (5.4) can be catalysed by trace constituents in the atmosphere. The idea is summed up in the reaction scheme

$$X + O_3 \rightarrow XO + O_2 \qquad (5.5)$$

$$XO + O \rightarrow X + O_2 \qquad (5.6)$$

$$\text{Net:} \quad O + O_3 \rightarrow O_2 + O_2$$

The reactive species X is regenerated in the second step, so that its abundance is not affected by its participation in odd-oxygen removal. Several species have been suggested for the catalytic 'X' in the atmosphere. The most important for the natural stratosphere are X = H and OH, X = NO, and X = Cl; the catalytic cycles are then said to involve HO_x, NO_x, and ClO_x. These cycles have activation energies for the individual steps that are lower than the activation energy of the direct $O + O_3$ reaction.

All three catalytic families, HO_x, NO_x, and ClO_x, appear to be present in the 'natural' atmosphere unpolluted by Man's activities. Precursors of the catalytic species have sources at the Earth's surface, and they have to be transported through the troposphere to the stratosphere. Amongst the species of importance are H_2O, CH_4, N_2O, and CH_3Cl, which are converted to the catalyst radicals in the stratosphere. One important route for conversion of the precursors involves excited atomic oxygen produced in the *ultraviolet* photolysis of ozone

$$O_3 + h\nu_{UV} \rightarrow O(^1D) + O_2(^1\Delta_g) \qquad (5.7)$$

The excess energy of the O* enables it to react with molecules such as H_2O

$$O(^1D) + H_2O \rightarrow OH + OH \qquad (5.8)$$

5.4 Generation of catalytic species via reaction with O*

Precursor	Products of reaction with O*
H_2O	OH + OH
CH_4	$OH + CH_3$
N_2O	NO + NO
CH_3Cl	$ClO + CH_3$

Note 5.4 shows some other products of the reaction of O*.

The stratosphere is very dry, probably because water from the troposphere has to pass through the 'cold trap' at the tropopause, and CH_4 constitutes more than a third of the total $[H_2O]+[CH_4]$ in the lower stratosphere. CH_4 is therefore an important source of OH, especially since the oxidation of the CH_3 radical (to CO) also yields two or three more odd-hydrogen species (e.g. H, OH, HO_2). Both N_2O and CH_4 are the result of biological activity (mostly microbial) on the Earth's surface. The main contribution to CH_3Cl is again biological, this time in the oceans, although burning of vegetation and some volcanic eruptions are additional sources.

Increased understanding of the role played by trace gases in determining atmospheric ozone concentrations has also led to an awareness that Man might inadvertently alter ozone concentrations by releasing catalytically active materials. Pollutants introduced into the stratosphere would have a lifetime for physical removal by transport of several years because of the vertical stability that results from the temperature inversion. They might therefore build up to globally damaging levels, reducing stratospheric ozone with biological consequences at ground level, such as increased incidence of skin cancer. Initial concern, in the early 1970s, centred on supersonic stratospheric transport (SST) aircraft, such as Concorde. Such aircraft could inject NO_x, produced from N_2 and O_2 in the high temperatures of the jet engines, directly into the stratosphere [Note 5.5]. Another source of increased stratospheric NO_x could be increased N_2O production in the biosphere resulting from intensive use of fertilizers. While perturbations due to the use of SSTs may be regarded as discretionary, agricultural use of fertilizers may be essential if populations continue to grow. Doubling the N_2O concentration is predicted to give a global ozone depletion of between 9 and 16%, although such large increases in N_2O are improbable in the near future. A more immediate problem seems to be the release of fluorinated chlorocarbons (CFCs), such as dichlorodifluoromethane, CF_2Cl_2 (CFC-12), and trichlorofluoromethane, $CFCl_3$ (CFC-11). The CFCs are extremely inert chemically, and are valuable as aerosol propellants, as refrigerants, as blowing agents for plastic foam production, and as solvents. The uses of the CFCs all lead ultimately to atmospheric release, and it appears that the quantities of CFCs in the troposphere are equal, within experimental error, to the total amount ever manufactured. Tropospheric inertness is thus confirmed, and lifetimes of up to hundreds of years are indicated. Only one escape route is available to the CFCs, and that is upward transport to the stratosphere. Sufficiently short-wavelength ultraviolet radiation penetrates to the stratosphere to photolyse the CFCs, the process involving the liberation of atomic chlorine, as exemplified for CFC-12

$$CF_2Cl_2 + h\nu \rightarrow CF_2Cl + Cl \qquad (5.9)$$

The Cl atoms can thus contribute to ozone destruction in the ClO_x cycle, and chlorine of Man-made origin now dominates over the natural CH_3Cl contribution. Many models predict significant ozone depletions from the release of CFCs [Note 5.6]. In recent years, a new phenomenon (the 'ozone hole') has been observed over the Antarctic, where extremely large depletions occur each year in October. The extent of the depletion seems to have been growing each year. It is now virtually certain that the 'hole' is associated with the increasing atmospheric load of CFCs.

5.5 Current numerical models indicate that ozone reductions due to SSTs are negligible, partly because the present fleet of aircraft is so small and partly because the aircraft fly low in the stratosphere where the NO_x cycle has relatively little effect on $[O_3]$. However, this issue is being reopened as discussions about a new generation of SSTs progress.

5.6 A feature of most predictions is that the depletions will get larger over one or more decades even if CFC release is curtailed now, and that the full recovery of ozone concentrations may take up to 100 years. There is now (1995) substantial evidence that global ozone levels are decreasing somewhat faster than expected. From 1988 onwards, a series of international protocols has restricted production of the CFCs, and they should have been phased out almost completely by 1996.

The troposphere

About 90% of the total atmospheric mass resides in the troposphere, and the bulk of the minor trace gas burden is found there also. The Earth's surface acts as the main source of the trace gases, although some NO_x and CO may be produced in thunderstorms. Hydroxyl radicals dominate the chemistry of the troposphere in the same way that oxygen atoms and ozone dominate stratospheric chemistry. Free-radical chain reactions initiated by OH oxidize H_2, CH_4 and other hydrocarbons, and CO, to H_2O and CO_2. The reactions thus constitute a low-temperature combustion system. The free-radical chain processes are photochemically driven, although stratospheric ozone limits the solar radiation at the Earth's surface to wavelengths longer than 290 nm. At these wavelengths, the most important photochemically active species are O_3, NO_2, and HCHO. All three can yield OH (or HO_2) indirectly, and thus initiate the oxidation chains. Ozone photolysis is, however, a critical step, since the other photolytic processes owe either their origin or their importance to it. Although only 10% of the total atmospheric ozone is found in the troposphere, all *primary* initiation of oxidation chains in the natural atmosphere depends on that ozone. Some ozone is transported to the troposphere from the stratospheric ozone layer, but a mechanism also exists for generation of ozone in the troposphere itself. If NO_2 is present, then NO_2 photolysis

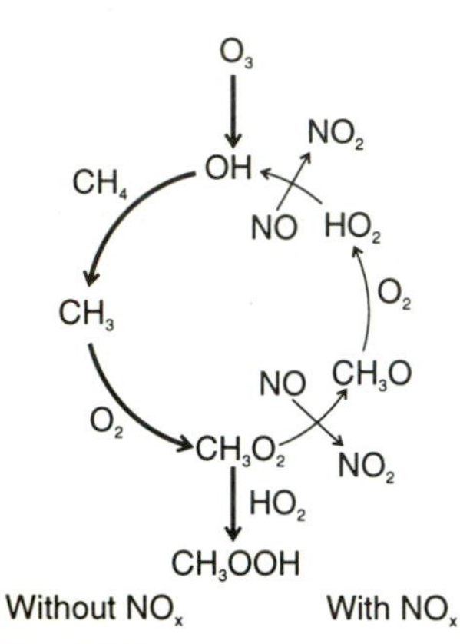

Fig. 5.4 Essential steps in tropospheric methane oxidation. The heavier arrows represent steps that can occur in the absence of NO_x. Only when NO_x is present can the loop be closed, OH be regenerated, and the process become cyclic.

$$NO_2 + h\nu\ (\lambda \leq 400\,\text{nm}) \rightarrow O + NO \qquad (5.10)$$

is a source of atomic oxygen, which can then form ozone in reaction (5.3)

$$O + O_2 + M \rightarrow O_3 + M \qquad (5.3)$$

The NO can itself be oxidized back to NO_2, as we shall see shortly, so that the formation of O_3 is not stoicheiometrically limited by the supply of NO_2 molecules initially present.

A first understanding of tropospheric photochemistry may be gained by considering methane as the only hydrocarbon present, and taking as our starting point the artificial situation where no CH_4 has yet been oxidized. Hydroxyl radicals must then be derived from ozone photolysis (at $\lambda \leq 310$ nm), in the way already described for the stratosphere in reactions (5.7) and (5.8). Attack of OH on CH_4 yields methyl radicals, and a sequence of oxidation steps ensues that we will follow, for the time being, to the formation of HCHO (formaldehyde)

$$OH + CH_4 \rightarrow CH_3 + H_2O \qquad (5.11)$$
$$CH_3 + O_2 + M \rightarrow CH_3O_2 + M \qquad (5.12)$$
$$CH_3O_2 + NO \rightarrow CH_3O + NO_2 \qquad (5.13)$$
$$CH_3O + O_2 \rightarrow HCHO + HO_2 \qquad (5.14)$$
$$HO_2 + NO \rightarrow OH + NO_2 \qquad (5.15)$$

5.7 Natural sources of NO_x include microbial actions in the soil, which produce NO as well as N_2O. Oxidation of biogenic NH_3, initiated by OH radicals, would be another significant source of NO_x. Lightning discharges appear to be responsible for less than 10% of the total NO_x budget.

Two very important features are displayed by this scheme. First, reactions (5.13) and (5.15) both provide a route for the oxidation of NO back to NO_2, and thus to a replenishment of tropospheric ozone through reactions (5.10) and (5.3). Secondly, the reactions as written are cyclic, the OH radical chain carrier being regenerated. The steps are illustrated in Fig. 5.4.

The oxidation steps for methane obviously have analogues for higher hydrocarbons, but in all cases the reactions depend on the switch between peroxy- (RO_2) and oxy- (RO) radicals in an interaction with NO. Oxides of nitrogen are therefore a central part of the oxidation scheme, because they both effect the switch and are the ultimate source of ozone, and thus of OH radicals. Figure 5.4 shows how NO is essential to the closing of the cycle.

Man's activities lead to the release of many kinds of pollutants to the troposphere; one form of pollution which is essentially photochemical in origin is the photochemical 'smog' from which many large cities suffer [Note 5.8]. The characteristic pollutants are ozone and nitrogen dioxide, together with a host of organic compounds. Concentrations of O_3 and NO_2 are so high that the ozone can easily be detected by smell, and a heavy load of particles leads to a brown haze in the air. Damage to materials such as rubber, damage to vegetation, reduction in visibility, and increased incidence of respiratory disease are recognized consequences of the pollution; the most immediately obvious effect of photochemical smog is eye irritation caused by substances such as formaldehyde, acrolein, and peroxyacetyl nitrate (PAN).

5.8 Los Angeles suffers particularly from photochemical smog. The pollution derives largely from automobile exhaust gases. Los Angeles has an exceptionally high traffic density, and sunshine is consistently intense, so that smog formation is favoured. In addition, the meteorological features of the Los Angeles basin, surrounded as it is by a ring of high mountains and the sea, lead to stagnation of the air and trapping of pollutants.

Nitric oxide, NO, is initially present, but is oxidized *after dawn* to nitrogen dioxide; ozone appears only after most of the nitric oxide has been oxidized. Small amounts of nitric oxide are known to be liberated, together with hydrocarbons, in automobile exhaust gases, and the presence of ultraviolet radiation is necessary to effect oxidation of NO and of hydrocarbons (see Fig. 5.5). The oxidation of both NO and the hydrocarbons is, in fact, a consequence of an exaggerated form of the chemistry already described for the natural troposphere.

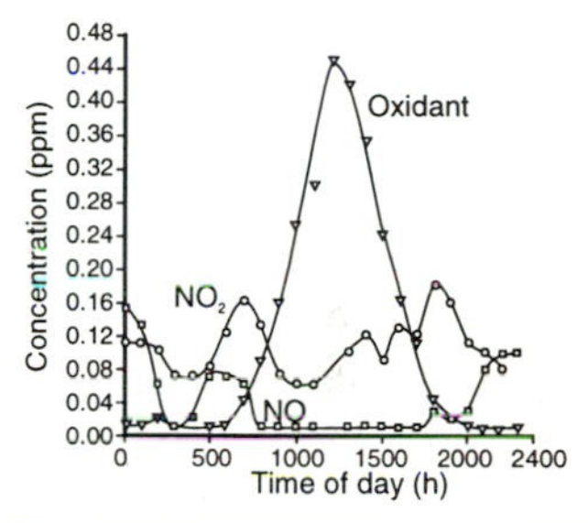

Fig. 5.5 Variations in concentrations of oxidant (mainly ozone) and oxides of nitrogen during the course of a smoggy day in Southern California.

Peroxyacetyl nitrate (PAN), $CH_3CO{\cdot}O_2{\cdot}NO_2$, formed by the addition of NO_2 to the acetyl peroxy radical, $CH_3CO{\cdot}O_2$, is an important component of photochemical smog, acting as an irritant of the respiratory system and the eyes, and being highly toxic to plants (phytotoxic).

Aerosols of particulate matter are found in many kinds of air pollution, including photochemical smog. The presence of suspended particles in the air leads to a serious reduction in visibility. Although the origin of the particulate matter in photochemical smog is not clear, it appears to involve the oxidative polymerization of hydrocarbons (possibly aromatic); laboratory studies have shown that aerosols can be formed by the irradiation of automobile exhaust gases. Aerosols are also produced in a form of natural air pollution, found in many parts of the world, but notably in the southwestern USA. A photochemically induced contamination of the atmosphere by particulate matter gives rise to a haze or smokiness over regions possessing high densities of trees such as pines or citrus fruit. Terpenes can be oxidized by ozone to give particulate matter [Note 5.9], and it seems that the atmospheric aerosols are formed in this way by reactions of terpenes liberated from the trees.

5.9 The formation of particulate matter discussed in the text can be demonstrated dramatically in the laboratory by squeezing a piece of orange peel near a flask of ozonized oxygen: a bluish cloud appears in the flask.

5.3 Photosynthesis

Photosynthesis is perhaps the most important of the many interesting photochemical processes known in biology; not only was the evolution of the Earth's atmosphere dependent on it, but also animal life derives energy from the Sun, via photosynthesis, by eating plants. It is estimated that the total mass of organic material produced by green plants during the biological history of the Earth represents 1% of the planet's mass, and that photosynthesis fixes annually the equivalent of ten times Mankind's energy consumption.

5.10 Our discussion of photosynthesis is confined to green plants, although it should be noted that there are certain other photosynthetic organisms (e.g. some bacteria) in which the essential photochemistry may be somewhat modified.

From the point of view of organic synthesis, the overall process consists of the formation of carbohydrates by the reduction of carbon dioxide

$$nCO_2 + nH_2O \xrightarrow{h\nu} (CH_2O)_n + nO_2 \qquad (5.16)$$

The essence of the process is the use of photochemical energy to split water and hence to reduce CO_2. Molecular oxygen is liberated in the reaction, although it appears at an earlier stage in the sequence of steps than the reduction of CO_2. True photochemical processes appear to produce compounds of high chemical potential, which can 'drive' the synthetic sequence from CO_2 to carbohydrate in a cyclic fashion.

Reaction (5.16) is thermodynamically very improbable in the dark. Production of one molecule of oxygen requires the transfer of four electrons, and four electrons are also needed to reduce one CO_2 molecule

$$2H_2O \rightarrow O_2 + 4e + 4H^+ \qquad (5.17)$$
$$4e + 4H^+ + CO_2 \rightarrow (CH_2O) + H_2O \qquad (5.18)$$

Reaction (5.16), with $n = 1$, is the sum of reactions (5.17) and (5.18), and it is evident that if each photon absorbed can lead to the transfer of one electron, then a minimum of four photons are needed for each CO_2 molecule converted. Experimental measurements of the quantum yield indicate that *eight* photons are needed in reality, suggesting that two photons are utilized for each electron transfer, and thus pointing to a two-step process with relatively long-lived intermediates connecting the steps. Other evidence, to be presented shortly, is in accord with this view.

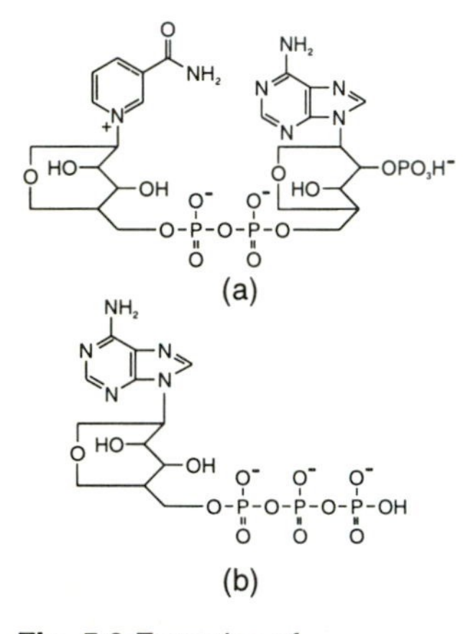

Fig. 5.6 Formulas of (a) nicotinamide adenine dinucleotide phosphate (NADP) and (b) adenosine triphosphate (ATP).

The essential feature of the actual carbon cycle is the input of energy by (and reducing power of) adenosine triphosphate (ATP) and the reduced form (NADPH) of nicotinamide adenine dinucleotide phosphate (NADP) (see Fig. 5.6). The synthetic carbon cycle can, in fact, be driven by ATP and NADPH in the presence of all initial enzymes and substrates, *but in the absence of light*. Thus it can be concluded that the ultimate result of the primary and secondary photochemical acts is the formation of ATP and NADPH, by the photophosphorylation of adenosine diphosphate (ADP) and the reduction of NADP

$$\text{ADP} + \text{inorganic phosphate} \xrightarrow{h\nu} \text{ATP} \qquad (5.19)$$
$$\text{NADP} + H_2O \xrightarrow{h\nu} \text{NADPH} + O_2 \qquad (5.20)$$

ATP is an 'energy-rich' phosphate used in many biochemical processes; the storage of chemical energy arises from the energy available in hydrolysis of ATP to ADP and H_3PO_4. Since reaction (5.19) can occur independently of CO_2 reduction, and in an anaerobic environment, the original development of the use of light by organisms may have been primarily for the storage of energy rather than for the synthesis of new organic matter. The development of photosynthesis proper would then have been a later evolutionary step.

The trapping and use of solar radiation depends on the presence of chlorophyll in the plant. Figure 5.7 shows the structure of the most ubiquitous chlorophyll, chlorophyll-a [Note 5.11]. Chlorophyll is an exceptionally efficient photosensitizer because of its ability to trap the energy of radiation and pass it on from one molecule to another until conditions are favourable for the sensitized reaction.

Reduction of NADP to NADPH (i.e. electron transfer from H_2O to NADP) requires more energy than is available in a single excited chlorophyll molecule, and it is necessary to postulate some kind of 'uphill' process, possibly involving several excited chlorophyll molecules. The actual energy requirement for reduction of 1 mole of CO_2 to carbohydrate is roughly 2.6 times the maximum available energy in a single excited chlorophyll molecule; with a quantum demand for the overall process of 8 (see p. 74), the efficiency of the multiple-quantum process must be as high as 33%. Some rather special and efficient 'uphill' mechanism is therefore in play in photosynthetic plants.

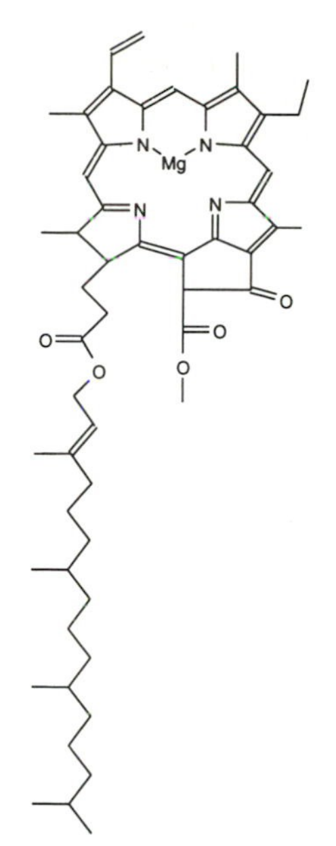

Fig. 5.7 Formula of chlorophyll-a.

Chlorophyll in the higher living plants is always to be found associated with lipoprotein membranes, which are organized into a highly ordered structure known as the chloroplast. The molecular arrangement in the chloroplast appears to be an essential component of the natural photosynthetic process. The chloroplast functions as a device both for the capture of light, and for the conversion of light energy to chemical energy. Several compounds other than chlorophyll are found in the chloroplast, including the auxiliary pigments (see Note 5.12), and carotenoid compounds which can both act as auxiliary pigments and protect chlorophyll against oxidative degradation [Note 2.31]. Also present are quinones (e.g. plastoquinone, α-tocopherol quinone, vitamin K) and proteins known as cytochromes that play roles in photosynthesis as important as those of the auxiliary pigments and carotenoids.

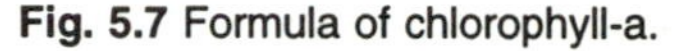

5.11 The resonance of the conjugated system in chlorophyll brings the optical absorption into the visible region of the spectrum, where the solar intensity is highest at ground level. The stability conferred by the porphyrin structure ensures that absorption of radiation is followed by energy transfer or radiative processes rather than by dissociation.

The membrane systems within the chloroplast are illustrated in Fig. 5.8: they seem to consist of a number of flattened sacs. Electrons may be transferred from one side of the membrane to the other, in such a direction that oxygen is released on the inside and reduction occurs on the outside. The number of chlorophyll molecules present in each chloroplast is directly related to the number of membrane surfaces, with roughly 10^9 chlorophyll molecules in a typical chloroplast. It seems that the pigment (mainly chlorophyll) molecules might be spread as monolayers on the membrane surfaces, maximizing the surface area of the pigment for light absorption and for energy transfer at specific sites on the membrane. An array of pigment molecules appears to be able to transfer energy to a single reaction centre, only one molecule needing to absorb radiation. Energy transfer, via the coulombic long-range mechanism (see Section 3.5), redistributes excitation from the primary absorber to other pigment molecules and ultimately to the reaction centre. In this elegant way, the available light is collected efficiently, without molecules of all components of the chemical system involved in the secondary processes having to be present for each individual light absorber. The complex array of pigment molecules thus functions as a *light-gathering antenna*.

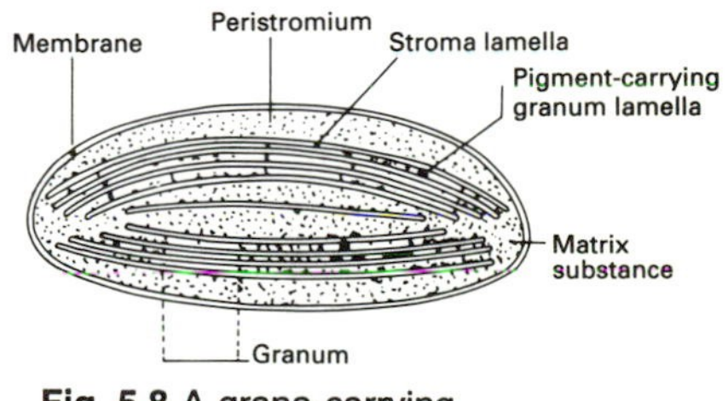

Fig. 5.8 A grana-carrying chloroplast.

Detailed examination of absorption spectra of chlorophyll reveals at least two forms of chlorophyll in the chloroplast, which may possibly be chlorophyll-a complexed to different proteins, or present as monomers and dimers. In the absence of more specific information about the two forms of

5.12 In photosynthetic organisms, the chlorophyll-a is usually accompanied by one or more auxiliary pigments whose function may be to absorb radiation at wavelengths where absorption by chlorophyll-a is weak; chlorophyll-b (chlorophyll-a with the 3-methyl group replaced by –CHO) and carotenes are probably the most important auxiliary absorbing pigments in the higher plants. The auxiliary pigments appear always to transfer their excitation energy to chlorophyll-a.

chlorophyll, they are known as 'pigment systems 1 and 2' (PS1 and PS2). PS2 absorbs at shorter wavelengths than PS1, probably because absorption occurs in an auxiliary pigment (e.g. chlorophyll-b in green plants): fluorescence studies indicate, however, that the excitation always resides on the chlorophyll-a and not on the auxiliary pigment.

The photosynthetic scheme consists of two partial oxidation and reduction reactions, electrons being transferred against the potential gradient by the absorption of light. Absorption of light by PS2 leads to the formation of a strong oxidant, which eventually yields molecular oxygen and a weak reductant. The process driven by PS1 generates a strong reductant and, concomitantly, a weak oxidant. NADPH is derived from the strong reductant finally produced, and ATP is generated in the reactions that link the two photosystems.

5.13 Efficient photosynthesis appears to require the simultaneous excitation of more than one photosynthetically active pigment, a result that suggests the possibility of two major processes in the energy-conversion reaction of photosynthesis.

The function of the chlorophyll in photosynthesis is now clear: under the influence of light, it can cause electron transfer and bring about oxidation–reduction changes. Two kinetic steps (electronic energy transfer and electron transfer) are of particular importance in photosynthesis, and each has associated with it a pigment–protein complex. True photochemical processes that involve electronically excited states are completed within about the first nanosecond after the absorption of light. The next stage in exploring photosynthesis will depend on separation of the pigment–protein complexes from the chloroplast, and it seems likely that a detailed description of the primary step in the complex molecular aggregate will soon further our understanding of what is one of the most important of all photochemical processes.

5.4 Vision

5.14 Vertebrates, molluscs, and arthropods have all developed well-formed eyes, although the anatomy and evolutionary development of vision in the three phyla are entirely different. It is therefore remarkable that the photochemistry of the visual process is nearly identical for the three types of eye.

Vision is stimulated by the photochemical transformation of a pigment containing a moiety related to retinal, which is the aldehyde derived from vitamin A (retinol). Extracts from the retina show the visual pigments to be compounds of a carotenoid substance with a protein. Rhodopsin, which is typical of the pigments, contains 11-*cis*-retinal (see Fig. 5.9) as the carotenoid chromophore, and the protein scotopsin. Animals derive their retinol from carotenoids of plant origin, and retinal is produced in the retina by enzymic oxidation of retinol. The opsins are proteins with relative molecular masses of about 40,000. Rhodopsin (bovine or ovine) has 348 amino acid residues grouped into seven mainly hydrophobic segments that pass between the two sides of the photoreceptor membrane.

The gross anatomy of the vertebrate eye — in particular, the system of lens and retina — is too well known to need description here. The receptors of the retina consist of 'rods' and 'cones': the former possess high sensitivity and are used at low light intensities, while the latter are less sensitive but may carry colour selectivity. In 1876, Böll discovered that the rose colour of the frog's retina faded in bright light. This bleaching of the so-called 'visual purple' demonstrated explicitly the occurrence of a photochemical reaction in vision. Subsequent studies showed that the bleaching is reversible if the retina is kept *in situ*: the reversibility was lost in solutions of visual purple extracted from the retina, even though the initial photobleaching still occurred. It is now recognized that the bleaching is too slow to be responsible for the sensory visual response, and that it is the end result of a sequence of reactions involved in nerve excitation.

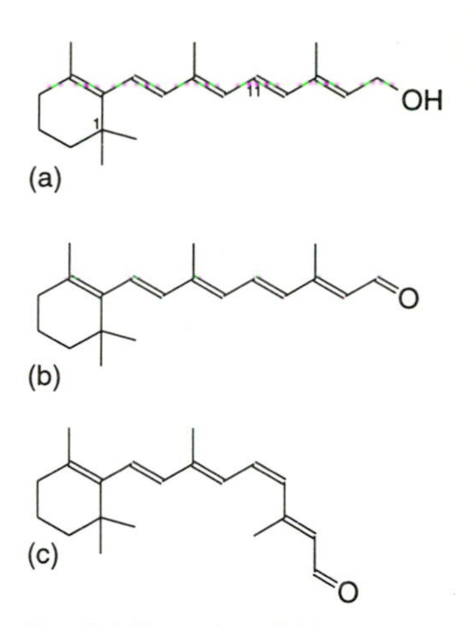

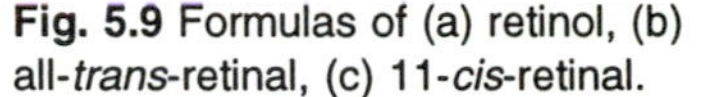

Fig. 5.9 Formulas of (a) retinol, (b) all-*trans*-retinal, (c) 11-*cis*-retinal.

Irradiation of rhodopsin leads to a series of conformational changes that are reflected in the appearance and disappearance of various intermediates of differing colour. Absorption of a photon leads to photoisomerization, and thus to strained structures and ultimately to cleavage of the protein–chromophore bond. Lowering of the energy of the excited state resulting from interaction of retinal with opsin leads to a red shift, and the stronger the interaction, the larger the shift. As the progressively more strained structures of lumirhodopsin and the metarhodopsins are formed, the shift becomes smaller and the absorption maximum moves towards the blue. In the case of bathorhodopsin, which absorbs at slightly longer wavelengths than rhodopsin, it may be that the ground state lies at a higher energy than the starting rhodopsin as a result of the strained geometry. The cycle is completed by the slow thermal isomerization of all-*trans*-retinal to the 11-*cis*-isomer that combines with opsin, and additional retinal can be supplied from vitamin A

$$\underset{red}{\text{Rh}} \xrightarrow{h\nu} \underset{red}{\text{batho-Rh}} \rightarrow \underset{orange}{\text{lumi-Rh}} \rightarrow \underset{orange/yellow}{\text{meta-Rh I/II}} \rightarrow \underset{colourless}{\text{retinal + opsin}} \quad (5.21)$$

Rh represents rhodopsin; all steps except the first one are thermal.

We now turn to a brief consideration of how the photochemical changes described so far become converted to an electrical impulse that stimulates the brain. There is evidence that a single quantum of radiation can stimulate a retinal rod. The absorption of one quantum does not, however, result in vision, and several quanta (between two and six is considered a reasonable estimate) must reach the same rod within a relatively short period. Even so, the process is remarkably efficient, and the energy of the ultimate reaction greatly exceeds that absorbed by the visual pigment. The absorption of light appears to initiate a chain reaction that derives its energy from metabolism, and visual excitation is a result of 'amplification' of the light signal received at the retina. The photoreceptor is the biological equivalent of a photomultiplier tube (see p. 15), which converts photons to an electrical signal with high gain and low noise. Both photoreceptor and photomultiplier achieve high gain in a cascade of amplifying stages. Visual pigments are integral membrane proteins that reside in the photoreceptor outer segment. Photoisomerization of retinal triggers a series of conformational changes in the attached protein that create or unveil an enzyme site. A cascade of enzymatic reactions follows, which ultimately produces a neural signal.

5.15 Colour vision is associated with the cones rather than with the rods. The pigment iodopsin has its absorption maximum at slightly longer wavelengths than the maximum of the absorption of the rhodopsin belonging to the rods. The cones are less sensitive than the rods, and the spectral response of the eye shifts towards the red in going from dim to bright light, as expected. Vertebrates appear to perceive colour through the operation of a three-colour system. Three different cone pigments seem to be implicated, with absorptions in the blue, the green, and the red wavelength regions.

6 Applied photochemistry

6.1 Introduction

Applications of photochemistry, ranging from photography to photomedicine, are of great importance to us. Only a few examples can be presented here, but those selected should illustrate the diversity of uses.

6.2 Photoimaging

The making of a more or less permanent record of light and shade by photography represents the best-known of all applied photochemical processes. Photography is one of a series of *photoimaging* techniques in which photons are used for the capture and replication of image information. Besides photography, other obvious large-scale applications of imaging include office copying and the preparation of various kinds of printing plates. If the image modifies the properties (e.g. the solubility) of a material used to protect some underlying medium, then subsequent treatment may allow the image to be transferred to the formerly protected surface. Such materials are known as *photoresists*, and are of enormous importance in the production of printing plates, integrated circuits and printed circuit boards for the electronics industry, the manufacture of small components such as electric razor foils and camera shutter blades, and many other applications besides [Note 6.1].

6.1 Considerable interest now attaches to photoimaging as a route to all-optical, as distinct from magnetic, memory storage devices. Applications of optical reading to video and audio ('compact disc') technologies are well established; optical read-write memories for computers are being developed.

Photoimaging can be divided into the three stages of capture, rendition, and readout. These stages are illustrated in Fig. 6.1 for a typical imaging system. The capture process in the example is photopolymerization; description of this important technique is deferred until Section 6.5. In general, image capture is the photochemical step, with image rendition consisting of the subsequent thermal reactions. Image readout then requires the development of the rendered image into a form that differentiates it physically from the unexposed background. Optical changes form the basis of ordinary photography; solubility changes can be used to generate three-dimensional relief patterns for printed circuits, printing plates, or three-dimensional topographic maps; wettability changes are used to produce lithographic printing plates, and enhanced tackiness can be used to pick up pigments to yield pigment-toned images in printing.

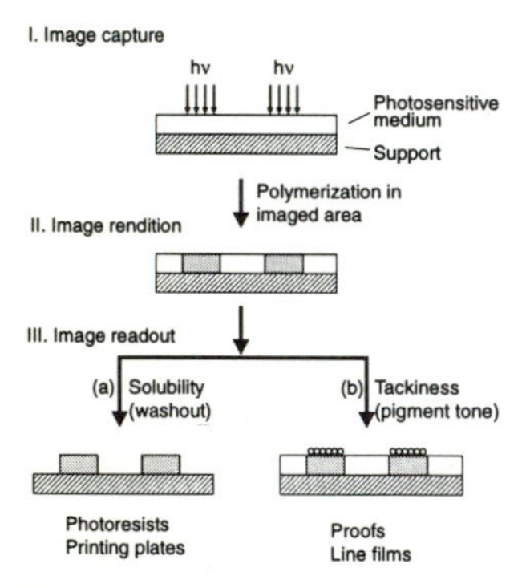

Fig. 6.1 Typical steps in photoimaging.

6.3 Photography

By far the most common form of the photographic process (both for monochrome and for colour photography) is based on silver halides as the photosensitive materials. Microcrystalline grains of silver halide suspended in gelatin are coated onto a suitable support (film, glass plate, paper, etc.) to form the light-sensitive 'emulsion'. Prolonged exposure to light causes darkening of the emulsion — the *print-out effect* — which X-ray powder patterns show clearly to be a result of metallic silver formation. Much shorter exposures produce a so-called *latent image* in the silver halide grains. This latent image may be turned into a visible silver deposit by a 'developer', which is a suitable reducing agent. All developers are, in fact, thermodynamically reducing towards silver halides, and the presence of the latent image seems to lead to an increased rate of reduction to metallic silver rather than to a change in the ultimate reducibility of the emulsion. Extended development of unexposed emulsion leads finally to darkening ('fog'), so that the discrimination by development between exposed and unexposed areas depends on the difference in reduction rate for the two areas [Note 6.2]

6.2 In normal practice, the developed image corresponding to exposed and unexposed areas is rendered permanent by 'fixing': the unexposed (and hence unaffected) grains of silver halide are dissolved in sodium thiosulphate solution immediately after development.

The latent image consists of metallic silver present in the halide grains, but at much lower concentrations than in the print-out image. The presence of the metallic silver atoms appears to lower the activation energy for the reduction reaction in development, and thus enhance the rate. Development, once it has been initiated at one site on a grain, proceeds with an increasing rate as more and more silver is formed, until the entire grain is developed. This autocatalytic activity of silver has been clearly demonstrated: a low concentration (about 10^{15} atoms cm^{-2}) of metallic silver evaporated onto a silver halide surface renders the surface 'developable'.

Silver halides show the phenomenon of photoconductivity, and it is believed that irradiation of silver halide raises photoelectrons from the valence band to the conduction band of the halide. The mechanism for the production of free silver then involves the migration of the photoelectrons and of interstitial silver ions to preferential sites in the halide; free silver atoms are formed by the combination of silver ions and electrons. The free silver so formed acts as an efficient trap for photoelectrons produced subsequently, so that further silver ions are discharged *near the same place* as the initial atom. Specks of silver grow, therefore, at the original preferential site. The positive 'holes' left behind by the electrons may have some mobility, and they can diffuse towards the surface of the halide grain to liberate free halogen.

6.3 The steps up to the formation of a two-atom speck are reversible, which is consistent with the experimental evidence for latent image stability only with aggregates of more than two atoms. A real crystal or grain of silver halide may possess chemical impurities and physical imperfections; in the case of emulsions, the gelatin may also enter into the photochemical process. These impurities seem to be important in practical photographic materials.

Spectral sensitization of silver halide emulsions can be achieved by adsorption of suitable dyes onto the halide grains. Sensitization is important, since it permits image formation by radiation of wavelength longer than that effective with unsensitized emulsions, and it offers an excellent example of a reaction photosensitized by energy or electron transfer. Spectral sensitization of photographic emulsions seems to have been the first recognized case of photosensitization (1873) (see Section 2.6).

Another use of dyes in silver halide photography is in producing images for colour photography. Here, much larger quantities of dye are used than in sensitization, and the dye becomes incorporated in the final image. A three-colour 'subtractive' process is used to give the entire range of visible hues. Three layers of emulsion are used, each being spectrally sensitized to a different spectral region (blue, green, and red), and possessing integral filter dyes if necessary. Exposure to a coloured image thus leads to the formation of different latent (silver) images in each layer. Although development involves reduction of the silver, as with black and white photography, in colour photography it is the consequent oxidation of the developer that is of interest. The oxidized developer reacts with a 'colour coupler', often incorporated in the emulsion layer, to produce an image dye.

6.4 Photochromism

6.4 Photochromic sunglasses and spectacles are familiar, and plastics incorporating a photochromic dye have also been used for aircraft windows that darken in bright sunlight but become lighter again under less intense radiation. Various kinds of data storage are possible, including image storage for uses like those of photography.

Silver halide photography involves the production of an essentially permanent optical effect by means of an ultimately irreversible photochemical process. A *reversible* light-induced colour change is referred to as *photochromism*. In photochromic systems, irradiation drastically alters the absorption spectrum; but when the irradiation source is removed, the system reverts to its original state. In some cases, the reversal can be brought about by light of a different wavelength. The visible effect often involves the appearance of colour in a previously colourless material, although changes in colour — for example from red to green — are also known.

The major difficulty in the use of materials incorporating photochromic materials is the rapid 'fatigue' exhibited by most known photochromic substances. Many of the photochromic systems reported are really able to undergo reversal only a limited number of times. Photochromism based on isomerization (see below) offers the best prospect of good fatigue characteristics, since, with alternative systems that involve bond cleavage, a very small lack of reversibility soon leads to chemical decomposition in side reactions.

6.5 Photochromic isomerization.

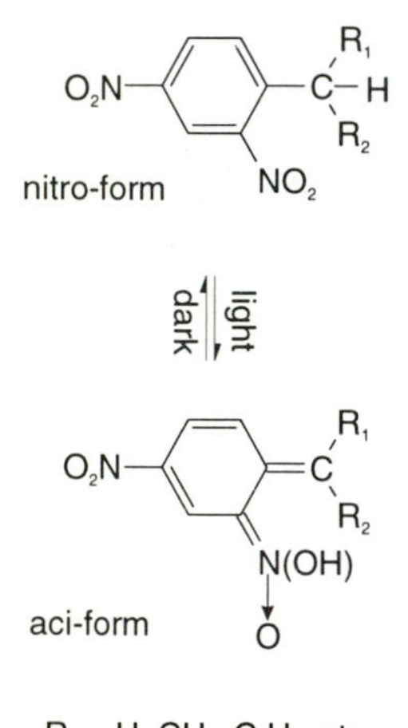

R_1 = H, CH_3, C_6H_5, etc.

R_2 = electron-withdrawing group

The main mechanisms responsible for photochromic behaviour are isomerization, dissociation, and charge-transfer or redox reactions. Many hundreds of specific photochromic substances are known, and one example must suffice to illustrate how photochromism arises. Many aromatic nitro-compounds exhibit photochromic isomerization: the process is believed to involve photoisomerization from the colourless nitro-form to the coloured aci-form [Note 6.5].

6.5 Photochemistry of polymers

Photopolymerization can be classified according to whether each increase in relative molecular mass requires its own photochemical activation step, or whether many (thermal) polymerization steps follow the absorption of a photon. *Photocrosslinking* falls into the first category, and involves formation of crosslinks between pre-existing polymer chains, while *photoinitiated polymerization* falls into the second.

Photopolymerization: imaging

The many uses of photoresists were listed briefly in Section 6.2. One important application is in the manufacture of electronic integrated circuits, where resists are used to define the doped regions on the silicon substrate that will form the resistors, capacitors, diodes, and transistors of the finished circuit, together with the metal conductors joining the components and the insulating and passivating layers [Note 6.6]. Photoresists are commonly based on photopolymer systems; those employed in the semiconductor industry are refined versions of photoresists used for making printing plates. Since crosslinked polymers are generally insoluble in any solvent, photocrosslinking might obviously form the basis of a photoresist system, and since the material that will be left behind after development by a solvent is that in the exposed regions, the resist will be *negative working* [Note 6.7].

6.6 In producing a complex circuit, there may be several tens of successive stages of imaging, followed by etching, doping, or other processing. Each stage must be lined up with an accuracy as good as a few hundred nanometres. Photographic methods are used to provide the required accuracy, although the use of ultraviolet radiation is being supplemented by shorter-wavelength X-rays or electron or ion beams, as even more components are packed into a small space (very large scale integration, VLSI).

One important photocrosslinking system involves the photolytic cycloaddition of pendant cinnamate groups incorporated in polymer chains. For example, on irradiation of poly(vinyl cinnamate), the cinnamoyl groups react to form a crosslinked structure

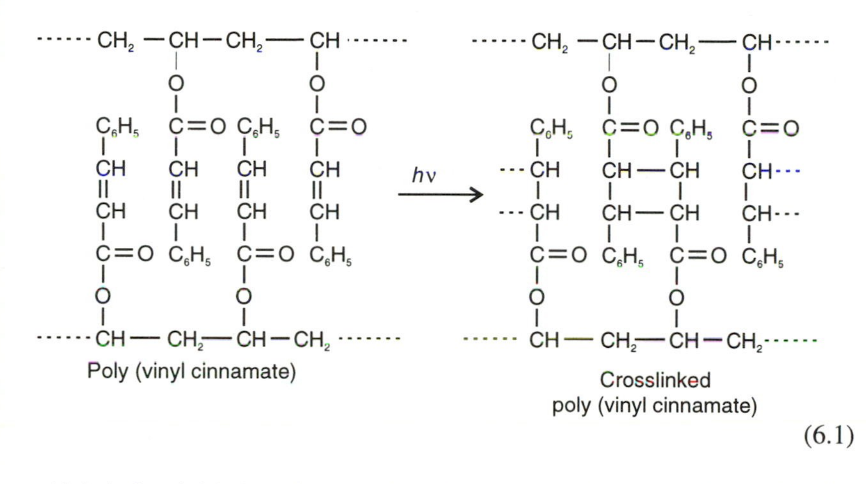

(6.1)

6.7 Resists in which the areas exposed to light are removed on development are referred to as *positive working*, since they leave behind the protective coating where the light was obscured by the transferred pattern. *Negative working* resists leave the coating which *has* been exposed to light.

which is insoluble in solvents. Dimerization can take place from either the singlet or the triplet excited states of the cinnamate chromophore. Only the *trans* isomer of the ester dimerizes, but the *cis* isomer is converted to the *trans* isomer by a photoisomerization reaction. Although the cinnamates only absorb radiation at wavelengths shorter than about 320 nm, triplet sensitizers such as 4,4′-bis(dimethylamino)benzophenone (Michler's ketone) can increase the sensitivity to near-uv and visible light by a factor of several hundred.

Photopolymerization: curing

Photoinitiation of polymerization has found little application for the bulk production of thermoplastic linear polymers, because satisfactory low-temperature *thermal* initiators are available. Rather, the major practical uses of photopolymerization are concerned with *in situ* polymerizations of relatively thin films of materials [Note 6.8]. The drying or hardening processes are generally referred to as *curing*, and the photochemical route offers considerable advantages over other methods. For example, many conventional coating techniques employ large quantities of solvents that play no part in the final cured coating, and whose purpose is to reduce the viscosity of the primary material to facilitate the coating operation; energy-inefficient

6.8 Apart from the various uses in imaging, photopolymerization of films is extremely valuable in applications ranging from the drying of decorative and protective coatings on a variety of substrates to the rapid and easily-controllable hardening of resins in dentistry.

6.9 Substituted acetophenones, such as benzoin esters, undergo α-cleavage (Norrish Type I: see p. 24) to yield initiating radicals. Improved performance is obtained by replacing the α-hydrogen by an alkoxy group, as in α,α-dimethoxyphenyl-acetophenone, which is photolysed according to the scheme

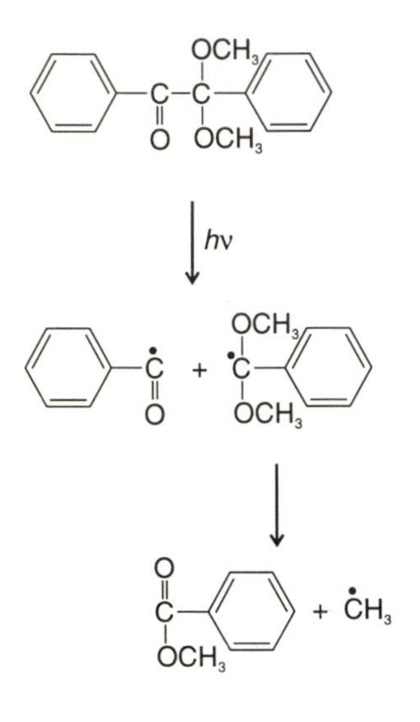

heating of the entire substrate may also be necessary. In contrast, photoinitiation of the cure works directly on the coating, so that the substrate need not be affected, and heating operations are not usually required. With high intensities, the cure can be completed in a fraction of a second.

Addition polymerization is used almost universally in photoinitiated curing operations, the monomers often containing either double bonds or strained rings. Most applications of photoinitiated polymerization involve a free-radical mechanism, with the monomer being based on acrylate esters (CH_2=CHCOOR). Acrylate groups are attached to resin types commonly used in coating technology (epoxides, urethanes, and polyesters). Polyfunctional reactive diluents, resulting from the reaction of polyols with acrylic acid, promote a more rapid cure and a more highly crosslinked coating. Successful photoinitiators have generally been aromatic carbonyl compounds, which have absorption spectra that match available ultraviolet sources.

Photodegradation and photostabilization

Our brief study of the photochemistry of polymers ends with two topics related to the durability of polymers in the outdoor environment. Most organic polymers undergo chemical change, or *photodegradation*, when exposed to visible or ultraviolet radiation, especially if atmospheric oxygen is present, and, as a result, the mechanical properties of the bulk polymer deteriorate. Durability is essential in some contexts, such as in the building or automotive industries, so that it is desirable to extend the usual lifetime by *photostabilization* of the material. On the other hand, there is also environmental concern with the persistence of agricultural plastics and of plastic packaging materials after disposal. Polymers may therefore be made deliberately light-sensitive; use of *photodegradable plastics* may allow articles such as plastic cups to be short lived, exposure to light reducing them to a fine powder, and thus 'naturally' disposing of abandoned waste.

Normal degradation is a light-initiated autoxidation process. Photostabilization of polymers must thus aim at reducing the rates of initiation or propagation of the chains, or possibly at increasing the rate of their termination. Reduction of residual impurities in the polymer would decrease the rate of initiation, and protection from oxygen would decrease the rate of propagation, but these two methods are rarely practicable.

An alternative way of reducing the rate of initiation is to prevent the absorption of light. Highly absorbent materials such as carbon-black are often used, and they confine photodegradation to the polymer surface. Reflecting substances such as the white oxides of zinc or titanium are used similarly. Quenchers may be used that prevent the relatively long-lived triplets of carbonyl compounds from participating in the secondary photoinitiation steps [Note 6.10]. Radical scavengers such as phenols, hydroquinones, and thiols can retard photodegradation by interfering with the propagation steps.

6.10 Orthohydroxybenzophenones make up one very useful class of stabilizer that operates by both screening and quenching. In addition, the hydroxybenzophenones seem able to react chemically with hydroperoxides, thus preventing the acceleration of autoxidation.

Deliberate incorporation of photoactive groups into a polymer can render it easily photodegradable, and thus confer environmentally desirable qualities. For example, copolymerization of ketonic species with hydrocarbons yields light-sensitive polymers. Photodegradation of the resulting material does not seem to involve radical-mediated photooxidation, but rather Norrish Type II scission of the polymer chain

$$-CH_2\underset{\displaystyle COR}{\underset{|}{C}}HCH_2CH_2- \xrightarrow{h\nu} -CH_2\underset{\displaystyle COR}{\underset{|}{C}}H_2 + CH_2{=}CH- \qquad (6.2)$$

Ideally, disposable agricultural film or packaging should undergo a sharp and controllable degradation initiated by exposure to ultraviolet radiation.

6.6 Solar energy storage

Solar energy is the ultimate source of the fossil fuels used by Man. Photosynthesis fixes about 2×10^{14} kg of carbon annually, or around ten times our present energy needs. Nevertheless, plants are not directly very useful energy sources, except perhaps in the case of wood. Rather, it is the decomposition and transformation over geological time spans that yields gas, oil, and coal. The recognition that the Earth's reserves of these fuels is not unlimited has stimulated photochemists to see how solar radiation might be used artificially. Aspects of such endeavours include: (i) the storage of energy for later release, as heat, light, or electricity; (ii) the immediate production of electrical power; and (iii) the formation, photochemically or photoelectrochemically, of products that would otherwise require consumption of power from other sources (such as production of Cl_2 from the Cl^- ion). In this section, we outline the principles underlying some of the purely photochemical methods being considered in the goal of solar energy conversion. Electrochemical methods are probably more promising commercially than the purely chemical ones, but are beyond the scope of this introductory text.

Certain features are common to all forms of photochemical solar energy utilization, and can conveniently be considered in their general forms here. It is apparent that any useful method of photochemical energy conversion must either use cheap and abundant raw materials [Note 6.11], perhaps to generate storable fuels, or must be capable of operating cyclically with regeneration of the energy-carrying material. The 'photochemical' processes all naturally involve the generation of electronically excited states, and the objectives are to use those excited states to form products (including electrons and ions) that are energy rich with respect to the reactants. From a thermodynamic viewpoint, energy beyond the energetic threshold for formation of the required products is 'wasted' since, in condensed-phase systems at least, excess energy will be degraded to heat in the absorbing medium. There is, however, another side to the question of excess energy. Whenever high-energy products are formed by an 'uphill' process, they are thermodynamically unstable with respect to the starting materials. The products can be harvested only if there are kinetic constraints to the reverse reaction. Two obvious constraints are that the reverse (exothermic) reaction must have an appreciable activation barrier,

6.11 Of possible consumable raw materials, water, carbon dioxide, and nitrogen must be amongst the cheapest and most abundant. The real prize is the photochemical fission of water, because hydrogen is a fuel that can provide pollution-free combustion, as well as being a feedstock for electricity-generating fuel cells.

which implies that the forward reaction also has a barrier greater than the endothermicity of reaction, or that the products must move apart fast and far enough not to re-encounter each other before they can be utilized, a situation again implying an input of energy greater than the thermodynamic threshold value. Purely homogeneous systems are at a serious disadvantage in this respect, in that the newly formed products are likely to be trapped within a solvent cage. In heterogeneous systems (or microheterogeneous ones such as micelles), the possibility arises that the products can be kept far enough apart to prevent the back reaction. We shall, however, only examine photochemistry in homogeneous solutions to provide a guide to what might occur in the more complex cases.

6.12 Absorption of light by one or other of the partners in a redox pair alters the redox potential because of the excitation of electrons. Excitation of the reducing partner makes electrons more readily accessible, and thus reduces the redox potential by an amount equivalent to the electronic excitation energy. Conversely, electronic excitation of the oxidizing partner increases the redox potential, because the electron hole left by the promoted electron makes the excited molecule a better electron acceptor (see Section 2.2).

A further critical factor in solar energy conversion is that the absorption spectrum must match the spectrum of the Sun's radiation incident on the Earth's surface. Optimum threshold absorption limits can be shown to lie in the range 1100–700 nm, with theoretical limiting thermodynamic efficiencies of around 30% for single-photon processes (see p. 75). It is worth pointing out that, although 700 nm is close to the threshold of absorption in green plants, the average efficiency of natural photosynthesis is nearly two orders of magnitude smaller than the thermodynamic limit.

Direct photolysis of water is not a candidate for solar energy conversion, since water does not absorb in the visible spectral region; fission to form H + OH as radical fragments has an energy threshold around $\lambda = 240$ nm, and even at that wavelength absorption is weak. An ionic redox mechanism, on the other hand, requires the transfer of four electrons which corresponds to a wavelength per photon of roughly 1000 nm, in the near infrared. Such a multiphoton redox fission of water thus seems promising, and the question is how to make it occur. The reactions involved would formally be

$$4H^+ + 4e \rightarrow 2H_2 \tag{6.3}$$
$$2H_2O \rightarrow O_2 + 4e + 4H^+ \tag{6.4}$$

One might imagine a redox pair that could oxidize water when unexcited, but that could reduce water when the reducing partner absorbed radiation [Note 6.12]. In an exactly similar way, a normally reducing pair might oxidize water on excitation of the oxidizing partner. These ideas are represented in Fig. 6.2.

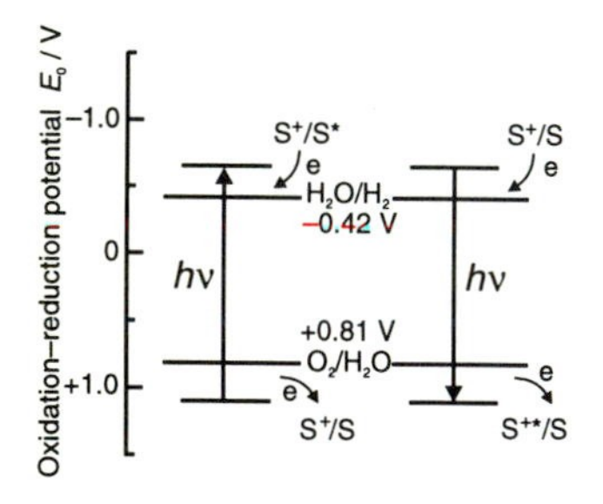

Fig. 6.2 The simultaneous photooxidation and photoreduction of water: redox schemes in which the light is absorbed (a) by the reductant and (b) by the oxidant partner of the couple S^+/S.

Unfortunately, real systems have yet to be found that operate quite so simply, and many ingenious indirect routes have been devised in order to overcome the various kinetic and energetic limitations. It has proved generally difficult to run both oxidation and reduction processes, reactions (6.3) and (6.4), with the same redox pair. Instead, many experiments have concentrated on one or other of the components, especially the less demanding reducing step. In that case, an external source of electrons must be found if the absorbing species is not to be consumed (i.e. if the redox system is to work catalytically). That source may be a chemical species that acts as a 'sacrificial donor'. For the analogous photooxidation process, a sacrificial acceptor is needed to remove electrons. Colloidal platinum can be used to trap electrons from one-electron reduced species, and so couple together the photochemistry and water reduction. Successful systems have also used an intermediate acceptor, or 'electron relay', between the photosensitizer and the platinum/water reactant. Using the symbols S for sensitizer, D for sacrificial donor, and A for intermediate acceptor, the general scheme is

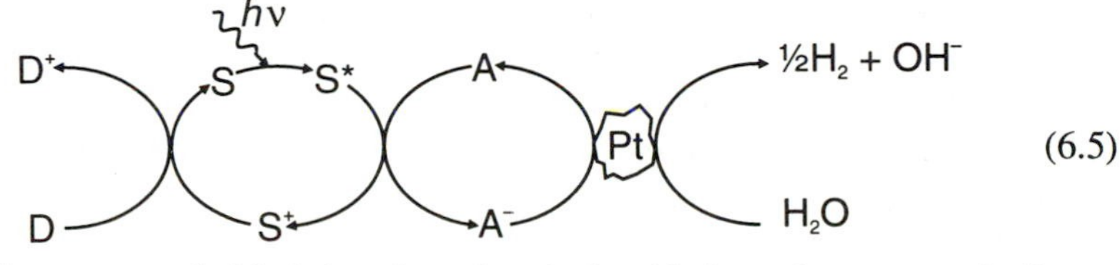

(6.5)

6.13 A redox couple much used in water splitting experiments involves a tris(bipyridyl) complex of ruthenium, represented by $Ru^{3+}(bipy)_3/Ru^{2+}(bipy)_3$. Two of the attractions of the ruthenium complex pair are the strong absorption and the relatively long lifetime of the Ru^{2+*} excited state that permits reaction with other partners. An effective intermediate acceptor is the compound methyl viologen (paraquat). The reduced form of the ruthenium couple is regenerated by a sacrificial donor also present, such as ethylenediaminetetraacetic acid (EDTA).

The concepts behind the photochemical oxidation of water are similar to those already explained, although greater problems are found in the practical implementation because of the need to transfer four electrons rather than two. Sacrificial acceptors are needed in this case, and persulphates have been found to be comparatively effective.

6.7 Photochemistry in synthesis

Photochemistry has important applications in synthesis in the chemical industry. A very few examples must suffice here to show the types of compound to which photochemical routes are suited. The main advantage of photochemical reactions in synthesis is that highly specific reactions can be brought about by light, to give products that would be difficult or impossible to form using thermal reactions. In the laboratory, photochemical methods are much used in the formation of four-membered rings by [2+2] inter- and intra-molecular cycloaddition and by electrocyclic ring closure of conjugated dienes (see pages 26–27 and 30–31). Photochemical techniques also offer a route to highly strained compounds whose thermodynamic instability disfavours thermal methods. Chapter 2 will have indicated the scope of photochemistry for laboratory synthesis.

We are more concerned here with the use of photochemistry in industrial synthesis. A photochemical process must obviously be superior in yield or purity of product in order for it to be competitive with alternative thermal methods. Reactions proceeding by a chain mechanism (often with radical chain carriers), in which the initiation step is photochemical, are particularly favourable candidates for industrial use. Indeed, we have already seen this application in connection with photopolymerization (Section 6.5). However, a photochemical reaction may even be economic if the quantum yield is low, so long as the chemical yield is higher than that available from thermal processes. In *fine* chemical manufacture, the use of light may represent only a small fraction of the total cost of the high-value product. Furthermore, the relatively small quantities of material involved mean that a batch process can often be used that is a scaled-up version of the laboratory method. Greater difficulties arise with the use of photochemistry in large-scale *heavy*-chemical manufacture, because the energy cost may now be a substantial element in the cost of the final product.

Chain reactions play a part in one of the most important applications: the chlorination of hydrocarbons. The process is illustrated by the reactions shown in Note 6.14. Kinetic chain lengths (see Section 4.3) can be as large as 1000: i.e. the overall quantum yield is very large, so that a relatively small radiation source can be used for high outputs of chlorinated material. In alkylated aromatic systems (e.g. toluene), photochlorination allows substitution in the alkyl group to the exclusion of chlorination of the aromatic ring. With benzene itself, addition occurs, to make hexachlorocyclohexane. The γ-isomer is a valuable biologically degradable insecticide, commonly known as gammexane or lindane.

6.14 The reactions involved in the chlorination of hydrocarbons are

$$Cl_2 + h\nu \rightarrow Cl + Cl$$
$$Cl + RH \rightarrow R + HCl$$
$$R + Cl_2 \rightarrow RCl + Cl$$

One example often quoted of a large-scale process that is commercially successful in spite of having a quantum yield of less than unity is the photooximation of cyclohexane to yield ε-caprolactam. Caprolactam is the cyclic amide that produces, by ring opening and addition, Nylon-6, which is a Nylon variant finding high-tonnage applications

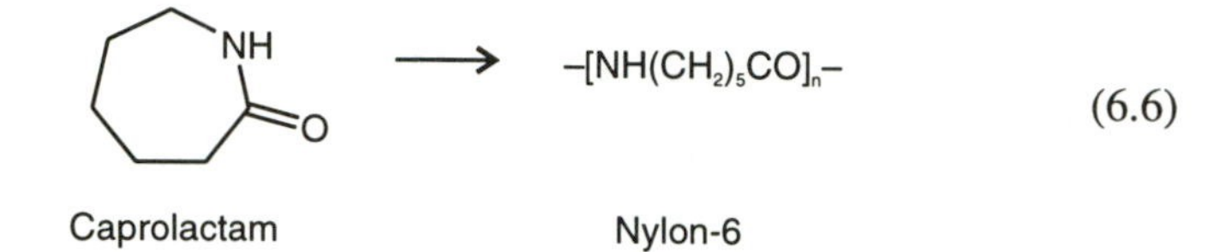

6.15 Photochemical route to caprolactam.

$$NOCl + h\nu \longrightarrow NO + Cl$$

Cl + (cyclohexane) ⟶ HCl + (cyclohexyl radical)

(cyclohexyl radical) + NO (or NOCl) ⟶ (nitrosocyclohexane) (+ Cl)

(nitrosocyclohexane) —HCl→ (cyclohexanone oxime, NOH) → (caprolactam, NH, O)

Caprolactam

In the manufacture of caprolactam, nitrosyl chloride (NOCl) is the photochemically active species used. Visible radiation will bring about fission of the weak Cl–NO bond, and the subsequent reactions of Cl and NO fragments with cyclohexane lead to the formation of an oxime and finally caprolactam [Note 6.15]. Hydrogen chloride is added to the nitrosating gas, and the ratio of NOCl to HCl is critical in determining the yield of the desired product. The cost of manufacturing caprolactam photochemically from 'inert' hydrocarbon compares favourably with that of producing the material by thermal reactions from starting materials such as phenol or toluene.

Photochemical production of fine chemicals is typified by the manufacture of vitamin D_3, a material used extensively in animal nutrition. 7-Dehydrocholesterol, made from cholesterol, undergoes electrocyclic ring opening on ultraviolet irradiation. The resulting triene, pre-vitamin D_3, is converted to vitamin D_3 itself by a thermally induced 1,7-hydrogen shift.

6.16 Nylon-12, $-[NH(CH_2)_{11}CO]_n-$, has important special uses for dimensionally stable articles, foodstuffs packaging, and coatings on metals. Its precursor, lauryl lactam (dodecanolactam) can also be produced by the photochemical route, this time using cyclododecane as the starting material.

6.8 Applications of lasers

Laser radiation can have particular properties such as monochromaticity, intensity, and coherence, that were indicated in Note 1.26. In suitable lasers, light pulses of very short duration can be generated. These special properties are valuable in a variety of applications, and some processes are only possible because of them. Examples of the uses of lasers range from laser welding to reading CD recordings, and from microsurgery to weapon aiming and guidance. It might be expected that lasers would also find applications in photochemistry itself. It is certainly the case that laser techniques have vastly improved the detail of the way in which photochemical processes can be probed. Lasers can provide the requisite light sources for highly selective excitation of specific energy levels, and pulsed lasers give access to time-resolved information (via laser flash photolysis and similar techniques) that was hitherto inaccessible. Furthermore, lasers are also at the heart of several modern and sensitive techniques for probing the excited states and fragments produced in photochemistry. The selectivity may be at the level of individual quantum states (rotational, vibrational, electronic) for both pump and probe aspects of an experiment. These are uses of lasers in experiments designed to *understand* photochemistry. Lasers also have applications in commercial procedures based on photochemical *processes.*

Lasers as light sources might be expected to find wide application in industrial synthesis. However, the high-power lasers needed are not yet available commercially, and laser methods are restricted in industry to selective separation of molecules at the atomic or molecular level. Photochemical separation of isotopes (mentioned on p. 37 in connection with IRMPD) is one such use. Laser isotope separation depends on the shifts in the optical absorption spectrum resulting from isotopic substitution. The highly monochromatic radiation from a laser can be used to excite only those molecules containing a particular isotope. Some means is then provided of removing or harvesting the excited species. For example, ^{35}Cl and ^{37}Cl may be separated by exciting ICl with radiation from a dye laser and allowing the excited molecules to react with 1,2-dibromoethene. Some 1,2-dichloroethene is formed, the simplified scheme

$$ICl + h\nu \rightarrow ICl^* \quad (6.7)$$

$$2ICl^* + C_2H_2Br_2 \rightarrow C_2H_2Cl_2 + 2IBr \quad (6.8)$$

indicating the nature of the interactions. If the laser is tuned to the absorption of $I^{37}Cl$ ($\lambda = 605.4$ nm), the resulting dichloroethene shows an enrichment in ^{37}Cl of nearly 20 over the natural isotopic abundance. A major stimulus to developing methods for laser isotope separation has been the need for the ^{235}U used in nuclear reactors. One enrichment scheme employs two-step photoionization of a molecular beam of uranium, in which the first laser selectively excites neutral U atoms, and the second photoionizes the excited (but not the ground state) uranium. The beam contains neutral molecules as well as ions, but the ions are preferentially of the 235-isotope. Electrostatic separation therefore enables the enriched sample to be collected.

6.9 Optical brighteners

Coloured fluorescent dyes and pigments find many non-scientific applications: they are used for the brilliant 'dayglo' paints, for textile dyes, and to obtain special theatrical effects. No application is so widespread, however, as the use of special fluorescent substances as *optical brighteners* or *bleaches* in the 'whiter-than-white' washing powders. The principle behind the operation of an optical bleach is that the substance should absorb in the ultraviolet and radiate in the visible region, so that the washed (white) textile apparently reflects more light than was incident upon it. A related large-scale application is in the optical bleaching of paper.

A substance that is to be suitable as an optical bleach must satisfy several stringent requirements. First, it must not absorb at all in the visible, since this would lead to coloration of the fabric, but must absorb strongly in the near ultraviolet, where there is still some intensity available from natural or artificial light sources. Secondly, the fluorescence must lie in the short-wavelength end of the visible spectrum, as otherwise the fluorescence would give an apparent undesirable yellowing to white fabric. Thirdly, the fluorescent substance must be photochemically stable, and it must not sensitize degradation or oxidation of the fibre material. Lastly, the substance must be soluble or dispersible in the aqueous detergent solution, but must be sufficiently strongly adsorbed by the textile fibres for an appreciable concentration to build up during washing and to remain after rinsing.

6.17 Cellulosic fibres are hydrophilic and swell in water so that the pores in the amorphous region grow to around 15–30 Å in diameter, large enough to admit the brightener molecules. Brightening of hydrophobic fibres (e.g. nylon, polyester, and acetate) takes place in a manner similar to the dyeing of these fibres, possibly involving the penetration of the molecules into canals between the fibre molecules, or, alternatively, as a result of actual solution of brightener in the solid fibre. Figure 6.3 shows some brighteners active for polyamide, polyester and acrylic fibres.

(R = alkyl)

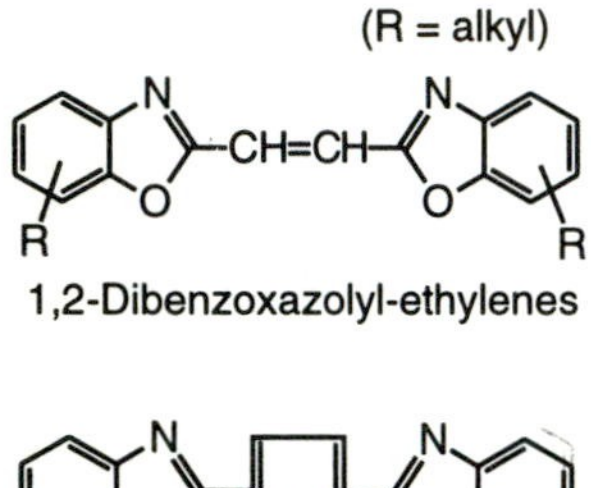

1,2-Dibenzoxazolyl-ethylenes

2,5-Dibenzoxazolyl-thiophenes

(X=Cl, H)

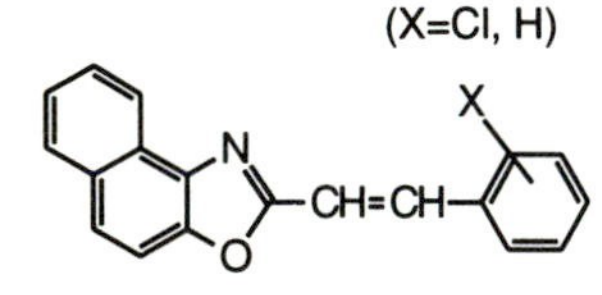

Styryl-naphthoxazoles

Fig. 6.3 Brighteners suitable for polyamide (e.g. Nylon), polyester (e.g. Terylene), and acrylic (e.g. Orlon) fibres.

6.10 Photomedicine

We conclude our survey of the applications of photochemistry with a very brief reminder that photochemistry has important uses in medicine. Ultraviolet radiation is employed in disinfection, sterilization, and the purification of water. Fluorescence is used diagnostically in dermatological and dental practice. Ultraviolet curing of dental resins was mentioned in Note 6.8, and lightweight orthopaedic casts based on photopolymerization have also been reported. *Phototherapy* is concerned with the treatment of disease. Minor skin ailments often respond well to exposure to ultraviolet radiation. Psoriasis is a serious skin condition, and is treated by *photochemotherapy*: exposure to ultraviolet radiation is augmented by use of a photosensitizing drug, such as 8-methoxypsoralen, taken by the patient some hours before the exposure. Sometimes the ultimate effect of the ultraviolet radiation is experienced in parts of the body different from those exposed. The production of vitamin D_3 described briefly at the end of Section 6.7 as an industrial synthesis, proceeds in the body in the same way, 7-dehydrocholesterol in the skin being converted photochemically to the vitamin. Normally, exposure to sunlight produces sufficient vitamin D_3, which is essential for healthy bone formation. In patients forced to remain indoors, supplementary exposure to artificial ultraviolet radiation may provide protection against the development of fragile bones and rickets. Several other examples of photochemotherapy have been described, and this rapidly developing field affords an excellent example of the chemical effects of light being harnessed to the benefit of Man, and also provides a suitably optimistic note on which to finish this introduction to the study of photochemistry.

Further reading

The first book is a classic detailed text that remains an important handbook for practising photochemists; the second two-volume work provides updated material. *Principles and applications of photochemistry* by one of the present authors is a rather more advanced treatment than this *Primer*. The next two books emphasize the way in which photochemistry is harnessed by nature and by Man. Barltrop and Coyle explain particularly clearly the behaviour of excited states and why it differs from that of ground states. The other texts are listed in alphabetical order, and represent just a selection of alternative, and more detailed, approaches to various aspects of our subject.

J. G. Calvert and J. N. Pitts, Jr. *Photochemistry*. Wiley, New York, 1966.

J. C. Scaiano, ed. *CRC handbook of organic photochemistry, Vols I, II.* CRC Press, Boca Raton, Fl, 1989.

R. P. Wayne. *Principles and applications of photochemistry*. Oxford University Press, Oxford, 1988.

J. D. Coyle, R. R. Hill and D. R. Roberts, eds. *Light, chemical change and life*. Open University, Milton Keynes, 1982.

P. Suppan. *Chemistry and light*. Royal Society of Chemistry, Cambridge, 1994.

J. A. Barltrop and J. D. Coyle. *Excited states in organic chemistry*. Wiley, London, 1975.

J. M. Coxon and B. Halton. *Organic photochemistry*. Cambridge University Press, Cambridge, 1987.

J. D. Coyle. *Introduction to organic photochemistry*. Wiley, Chichester, 1986.

A. Gilbert and J. E. Baggott. *Essentials of molecular photochemistry*. Blackwell Scientific, Oxford, 1991.

W. M. Horspool and D. Armesto. *Organic photochemistry : a comprehensive treatment*. Ellis Horwood, New York, London, 1992.

H. Okabe. *Photochemistry of small molecules*. Wiley, New York, 1978.

R. Schinke. *Photodissociation dynamics*. Cambridge University Press, Cambridge, 1993.

N. J. Turro. *Modern molecular photochemistry*. University Science Books, Mill Valley, Ca., 1991.

Index